Nuclear Safeguards Analysis

Nondestructive and Analytical Chemical Techniques

Nuclear Safeguards Analysis

Nondestructive and Analytical Chemical Techniques

E. **Arnold Hakkila,** EDITOR

Los Alamos Scientific Laboratory

Based on a symposium sponsored
by the Division of Nuclear
Chemistry and Technology at the
175th Meeting of the American
Chemical Society, Anaheim, CA,
March 13–17, 1978.

ACS SYMPOSIUM SERIES 79

AMERICAN CHEMICAL SOCIETY
WASHINGTON, D. C.   1978

Library of Congress CIP Data

Nuclear safeguards analysis.
  (ACS symposium series; 79 ISSN 0097-6156)

  Bibliography: p. 214 + x.
  Includes index.

  1. Nuclear fuels—Testing—Congresses. 2. Nu-
clear fuels—Analysis—Congresses. 3. Atomic energy
industries—Security measures—Congresses.
  I. Hakkila, E. Arnold, 1931–   . II. American
Chemical Society. Division of Nuclear Chemistry and
Technology. II. Series: American Chemical Society.
ACS symposium series; 79.

TK9360.N854   621.48'35   78-12706
ISBN 0–8412–0449–7   ACSMC8 79 1–214 1978

ACS Symposium Series

Robert F. Gould, *Editor*

2602

FOREWORD

The ACS SYMPOSIUM SERIES was founded in 1974 to provide a medium for publishing symposia quickly in book form. The format of the Series parallels that of the continuing ADVANCES IN CHEMISTRY SERIES except that in order to save time the papers are not typeset but are reproduced as they are submitted by the authors in camera-ready form. Papers are reviewed under the supervision of the Editors with the assistance of the Series Advisory Board and are selected to maintain the integrity of the symposia; however, verbatim reproductions of previously published papers are not accepted. Both reviews and reports of research are acceptable since symposia may embrace both types of presentation.

CONTENTS

PREFACE

Nuclear safeguards is becoming an increasingly important factor in the public acceptance of nuclear energy, particularly at the end of the nuclear fuel cycle where strategic nuclear materials (SNM) in the form of high-purity plutonium and enriched uranium are available in concentrated forms that are attractive to a potential divertor. Effective safeguarding of nuclear materials relies on a combination of physical security, materials control, and materials accountability.

No materials control or accountability system can be considered adequate without suitable measurement techniques. Measurement methodology must provide timely, rapid, precise, and accurate means of determining location and quantity of SNM. In-line or at-line measurement techniques rely heavily on nondestructive analysis (NDA), including x-ray, gamma-ray, or alpha-particle emission or absorption, active or passive neutron interrogation, or calorimetry. These NDA methods are complemented by conventional analytical chemical methods that, although are often not as rapid, are capable of providing improved precision and accuracy. Finally, it must be recognized that no measurement system is complete without a standards program whereby data can be correlated to precisely known reference materials which can be traced to a national standards program.

This symposium on nondestructive and analytical chemical techniques in nuclear safeguards was organized to review some of the methodology required for an effective measurement program. The overall safeguards program in the national laboratories is directed from the Department of Energy, Office of Safeguards and Security (DOE–OSS). Chapter One reviews safeguards needs as assessed by DOE–OSS while the following two chapters review the standard-materials programs operated by the New Brunswick Laboratory and the National Bureau of Standards, respectively. Chapters Four and Five discuss the development of data evaluation methodology for diversion detection in dynamic materials accounting, which is a key element in future safeguards systems, and the nonlinear curve fitting techniques that allow for both standards and measurement uncertainties.

The key input–accountability measurement for nuclear fuel reprocessing plants will be at the accountability tank. This measurement is correlated with plant output measurements and with reactor operating data, and is discussed in Chapters Six and Seven. The next four chapters describe some of the NDA instrumentation methods being investigated for on-line applications, while the final chapter (which was not presented at the symposium) describes an accountability system at an operating reprocessing plant. It includes a description of some of the measurement techniques and presents typical data that have been obtained.

I want to express my thanks to the participants for their contributions toward making the symposium a success, and to Dr. Clemens Auerbach of Brookhaven National Laboratory for chairing one of the sessions.

Safeguards Systems Group Q-4   E. ARNOLD HAKKILA
Los Alamos Scientific Laboratory
Los Alamos, NM
June 20, 1978

Safeguards Needs in the Measurement Area: the Realm of Measurements

GLENN HAMMOND

Office of Safeguards and Security, U.S. Department of Energy (DOE), Washington, DC 20545

CLEMENS AUERBACH

Department of Nuclear Energy, Brookhaven National Laboratory, Upton, NY 11973

The ACS meeting and its program on "Nondestructive and Analytical Chemical Techniques for Nuclear Safeguards" provides a timely forum for permitting all of us from the various measurement areas in the nuclear field to participate and contribute to the vital safeguards challenges. We wish to jointly share some of our ideas on measurements for safeguards, their evolution, and highlights and objectives of the emerging measurement advances.

DOE's safeguards program relates to all its nuclear materials and facilities, but concentrates on the more readily usable forms of fissile or special nuclear material (SNM) – plutonium, uranium enriched in ^{235}U and ^{233}U. It supports the development of safeguards concepts for new power reactor designs and related fuel cycle facilities. The general objective of the nation's safeguards program is to prevent successful malevolent acts involving special nuclear material and facilities. The term "safeguards" then is used in a broad sense to include physical protection and materials control measures to deter and detect theft and to provide a monitoring and accountability capability for SNM flow streams, transactions and inventories. In addition, DOE considers U. S. national security to dictate that nuclear materials and facilities wherever they appear in the world, should be protected against malevolent action as well as safeguarded internationally against nuclear proliferation.

Reliable materials control and accountability include the need for (1) timely characterization of the material to determine the intensity of protection needed and quantitative determination of what, where and how much material is being protected (or requires protection); (2) rapid detection and localization of a loss and backup to physical protection; (3) effective means for investigation and, if necessary, to initiate actions for recovery, and (4) frequent testing for credible confirmatory assessment that the protection and control systems are working properly and have not been

circumvented. The continuous monitoring of material to
meet these needs will, in addition, help meet plant requirements for process and quality control, materials management,
criticality control and health and safety.

     To give you a picture of the important role that measure-
ments play in safeguards and the development necessary for
successful implementation, we would like to first review
briefly the history of nuclear materials measurements;
second, the rationale for reliable measurement in materials
control and accountability; third, types of measurements of
nuclear materials including traditional chemical and isotopic
analyses, and the newer non-destructive techniques; fourth,
a national nuclear standards and measurement assurance
program; and finally, the challenges we see in accomplishing
the various tasks involved.

BACKGROUND AND EVOLUTION

     The control and measurements of nuclear materials are
not new. The nuclear materials produced at Oak Ridge and
Hanford in the early 1940's were guarded carefully because
of their extremely limited quantities and very sensitive
potential military application. The years just preceding
World War II were marked by a dramatic evolution of analytical
chemistry of nuclear materials as a science, drawing freely
on developments in physical chemistry and other related dis-
ciplines. Since U. S. scientists were in the forefront of
some of these developments, the Manhattan Project was able,
just as in a number of other areas, to profit not only
from the new advances but also from direct collaboration
with the key scientists responsible for them. With the aid
of individuals like N. H. Furman of Princeton University,
C. F. Metz of the then newly established Los Alamos Scientific
Laboratory and many others, the analytical procedures developed in those days went far beyond serving their immediate
purpose. In conjunction with new developments in radio-
chemistry, microchemistry and separation techniques, these
procedures set a trend for analytical techniques which has
not been surpassed since, and indeed gave major impetus to
the major advances in actinide chemistry. In similar
fashion, mass spectrometric techniques for measuring the
isotopic composition of uranium were developed in response
to the demands of the Manhattan Project, to a point where
they could eventually be adopted by the civilian nuclear
industry without much fundamental change.

     During the early years, the relatively small physical
inventories were difficult-to-measure however, using available manual techniques which were tedious and time-consuming.
Methods and techniques for chemical and isotopic analyses
were still being developed for the new materials. There

was virtually no instrumentation to reduce analysts time,
few standard reference materials or standard methodology,
and few laboratories for intercomparison of results. In
fact, in many cases, the analysis was performed by the analyst who had just developed the method, and staff who were
familiar with statistical methodology were not always available to compare and review results.

Shortly after the war, the Atomic Energy Act placed
responsibility for this new energy source in the hands of
a civilian agency. The nuclear material processes and
operations were still being developed and lacked efficiency.
The material was expensive to produce, and emphasis was
placed on financial responsibility for control.

In subsequent years, priorities -- technical, economic,
and political -- related to nuclear energy changed. Legis-
lation, the 1954 revision of the Atomic Energy Act and the
Private Ownership Act of 1964 (Public Law 88-489) permitted
expansion of the use of nuclear energy and related materials.
Two years later in 1966, federal regulations were adopted
which placed a specific obligation on the domestic private
industrial sector to safeguard SNM. When international
terrorism escalated in the early 1970's, nuclear materials
and related facilities, at home and abroad, were recognized
as possible targets for terrorist purposes because of the
potential for extensive malevolent use and the growing anti-
nuclear interests. The concept of balanced and integrated
systems was recognized as a means to improve effectiveness
of safeguards.

These developments led to, among other things, evolutionary changes in chemical and isotopic measurement methods
along the lines of increasing reliability and speed using
standards and automation. The AEC and its successor organizations (ERDA, NRC, DOE) have consistently played a major
role in supporting these activities, with the result that
measurement techniques at the U. S. Government-owned laboratories have become unique in terms of size, versatility
and sophistication.

International safeguards, as carried out by the
International Atomic Energy Agency (IAEA), places reliance
on materials measurements in accountability systems. Significance is attached to quantities of nuclear materials that
could be used by a country as part of a nuclear explosive
device. DOE, in cooperation with other U. S. Government
agencies, including the State Department, the Arms Control
and Disarmament Agency (ACDA), and the Nuclear Regulatory
Commission (NRC) provides both safeguards experts and equipment to assist the IAEA. Implementation includes directing
a technology base towards answering technical questions
posed by U. S. non-proliferation initiatives and by U. S.
participation in the International Nuclear Fuel Cycle

Evaluation (INFCE) program being conducted by the IAEA.
Today, national and international safeguards concerns are
being addressed in the development of concepts and supporting measurement technology for safeguard systems for spent
fuel storage, uranium enrichment, chemical reprocessing or
coprocessing and proliferation-resistant alternative fuel
cycles. These efforts include major support to a national
Nonproliferation Alternative Systems Assessment Program
(NASAP).

   Today, fuel cycle alternative requires a comprehensive
study and evaluation of measurement methods and instruments
for the range of material forms and compositions which are
characteristic of the related processes. Accuracy, precision
and operational features are required for on-line and at-line
instrumentation to optimize materials management, control
and accounting systems.

RATIONALE FOR AN EFFECTIVE SAFEGUARDS MEASUREMENT SYSTEM

   Measurements and measurement quality assurance programs
are vital to materials control and accountabilty safeguards
systems. Material balance accounting is drawn around a
plant and several major portions of the plant processes by
adding all measured receipts to the initial measured inventory
and subtracting all measured removals from the final
measured inventory. Measurements establish the quantities
of nuclear material in each custodial area and a facility
as a whole as one of a number of safeguards subsystems
contributing to the desired capability to localize losses
and in generating and assessing safeguard alarms. Of
course, appropriate checks and balances are required to
detect mistakes and protect the material accounting system
from fradulent source data; and a strict measurement quality
assurance program is necessary to ensure the accurate
calibration of the measurement systems and the reproducibility of the measurements.

   As part of the safeguards system, nuclear facilities
are required to establish and report, on a regular basis,
material balances based on these measured values. Regulations to this effect have been promulgated by DOE and
NRC. These regulations center on the concept of inventory
differences (ID), previously known as Material Unaccounted
For (MUF), and defined by the expression

$$ID = BI + A \ -EI \ -R \ :$$

where BI = beginning inventory
      A = additions to inventory since the last
          physical inventory
      EI = ending inventory

R =   removals from inventory since the last physical
        inventory.

   If all uncertainties, biases, transcription errors,
process holdups, unmeasured losses, etc. are properly
accounted for, then in the absence of theft or diversion ID
should be zero. The regulations stipulate that ID's exceeding
predetermined limits of error (LEID) shall be viewed as the
result of possible theft or diversion of nuclear material
and appropriate action taken. Limits on how large the
measurement uncertainty may be, based either on a fixed
amount or on a ratio of throughput, are determined by statistical means. The statistical means and appropriate mathematical modeling techniques have recently received additional
interests by DOE and some of its contractors. The goal is
to develop statistical error propagation methodology which
will permit evaluation of appropriate LEID's from facility
measurement control data. The goal includes a graded approach
whereby the LEID values will reflect the strategic significance of a given nuclear material stream (flow or inventory).
Much has been written on the use of mathematical statistics
in evaluating the complex problems associated with safeguards
systems. We note in particular the work by John Jaech,
"Statistical Methods in Nuclear Material Control" (1).

THE REALM OF MEASUREMENTS

   The realm of measurements for safeguards includes a
variety of techniques required for characterizing and
determining nuclear material quantities in feed, process,
product and waste streams; for standards and measurement
controls, performance evaluation and system optimization;
and for independent verification by a safeguards inspector-
ate.
   An effective safeguards measurement system must combine
the elements of versatility, reliability and timeliness.
Streams to be measured include materials ranging from essentially pure uranium and plutonium compounds, which are
relatively easy to sample and dissolve, to heterogeneous
and intractable solid waste generated in the course of
processing operations. This may include such diverse
items as used casting crucibles, contaminated paper, rags,
rubber gloves, floor and hood sweepings, etc. Each stream
must be measured with an accuracy and precision commensurate
to the contribution which the stream makes to the overall
nuclear material balance. The guiding principle is the
establishment of a fully measured material balance within
predetermined limits of error.
   To satisfy these wide-ranging and sometimes conflicting
demands, a systematic and judicious choice must be made

between methods which may be considered in three inter-
related areas: (1) bulk measurements which are directed
to total volume and flow rate, gross and net weights, and
total piece count; (2) sampling which is directed to obtaining a representative and tractable portion of a total
batch under consideration; and (3) analytical determinations
which are directed to specific characteristics (chemical,
physical, nuclear) of the material under consideration.
Analytical measurements then may be categorized broadly
into chemical and nondestructive methods. Chemical methods,
in the present context, are based on sampling followed by
laboratory measurements of either concentration or isotopic
composition of SNM. Combined with appropriate bulk measurements these methods yield the total quantity of SNM in a
given flow stream or inventory stratum. Nondestructive
analysis (NDA) is based on the nuclear properties of uranium
and plutonium; these properties are used to measure the
SNM content of material which cannot be sampled in representative fashion or which does not easily yield to dissolution.
        It is now recognized that a truly effective safeguards
measurement system must make concerted use of both chemical
and nondestructive methods. Accordingly, the thrust of
recent DOE-sponsored research and development has been
towards potential solutions which incorporate the most
desirable aspects of both approaches. Work at Los Alamos
Scientific Laboratory, Lawrence Livermore Laboratory, New
Brunswick Laboratory (now located at Argonne, Illinois)
and at other laboratories and facilities, is directed at
making chemical methods both more timely by way of automation,
and more responsive to non-homogeneous or otherwise intractable materials, even to the extent of incorporating some
aspects of NDA methodology. At the same time, advances in
electronics and detector capability combine to make possible
increasingly sophisticated NDA approaches, in terms of both
versatility and accuracy. A significant aspect of these
developments is the close collaboration between DOE con-
tractor laboratories and facilities abroad, notably in the
Federal Republic of Germany and other members of the
European community and the International Atomic Energy
Agency (IAEA). Some of these developments will be covered
in detail by other participants in this Symposium.
        Other than a few exceptions, chemical methods in use
today are based in essence on developments which took place
during the Manhattan Project and have not changed significantly in terms of the general principles involved. In
1963, the AEC with the assistance of an Advisory Committee
for Standard Reference Materials and Methods of Measurement reviewed, evaluated, and published "Selected Measurement Methods for Plutonium and Uranium in the Nuclear

Fuel Cycle" (2). The publication was revised in 1972 to
recognize intervening improvements (3).

These wet analytical methods are in existence at
nuclear facilities to measure uranium and plutonium in a
variety of materials -- metal, alloys, salts and oxides.
Much of DOE's work related to the improvement and automation
of analyticl methods to reduce uncertainties in inventories
or materials balance control is being carried out at the
New Brunswick Laboratory (NBL) and the Lawrence Livermore
Laboratory (LLL); and at the Los Alamos Scientific Laboratory
(LASL) related to fast dissolution methods for refractory
nuclear materials, and the testing of an inexpensive mass
spectrometer for in-plant inspection use.

The Davies-Gray method which is used for determining
uranium has been the subject of extensive development work
both at NBL and LLL. The original method is based on the
reduction of $U(VI)$ to $U(IV)$ with $Fe(II)$ in H_3PO_4 solution,
followed by oxidation of excess $Fe(II)$ with HNO_3 in the
presence of a $Mo(VI)$ catalyst and titration with $K_2Cr_2O_7$
to a colorimetric (visual) end point. The method was
improved and refined at NBL by the addition of $V(IV)$ to the
solution to markedly speed up the attainment of equilibrium,
which allowed the use of potentiometic end-point detection.
These efforts have resulted in a fully automatic uranium
titration system, developed by LLL and delivered to the new
NBL site at Argonne National Laboratory (ANL) in 1976.
This system is being tested currently for non-irradiated
uranium, including uranium alloys and scrap. Some 44 samples
can be analyzed in an 8 hour day with a relative standard
deviation of about 0.1%, using 20-150 mg samples. Complete
fault and malfunction detection hardware and software are
used to permit unattended operation. An attractive feature
of this system is that it automatically shuts down should
data not match standard uranium values. Analytical results
are calculated and printed on a hard copy minicomputer.

The Analytical Chemistry Group at LASL has been responsible for a large number of important developments in the safeguards measurement area. One of the more interesting ones
concerns an overall analytical system for scrap and other
hard-to-dissolve material. The system is depicted schematically in Figure 1, which demonstrates how an overall error
of less than 1.5% may be attained even though about 10%
of the sample cannot be dissolved. A high pressure dissolution technique is employed, and subsequent automated
chemical analysis is performed by a spectrophotometer
system which was developed at LASL and has since been
refined. The instrument incorporates a solvent extraction
system and dual filters which enable sequential analysis
of U and Pu in the same solution. The instrument can
accommodate samples in the milligram to submilligram range.

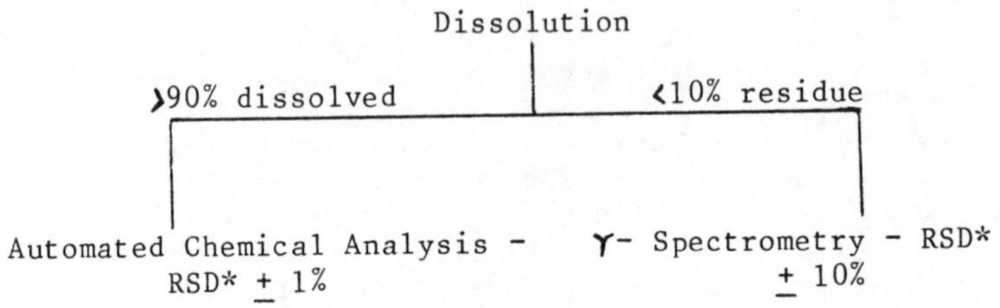

Dissolution

〉90% dissolved   〈10% residue

Automated Chemical Analysis - Υ- Spectrometry - RSD*
RSD* $\pm$ 1% $\pm$ 10%

Overall error = $\left[(10 \times 0.1)\right]^2 + \left[(1 \times 0.9)\right]^{2\,1/2} - 1.3_4\%$

*Relative standard deviation

Figure 1. Overall analytical system for scrap and other hard-to-dissolve materials

These chemical methods are capable of the same high
accuracy and precision (0.1% or better) characteristics of all "classical" analytical techniques. Apart
from their lack of timeliness, the chief limitation of
chemical methods is that many streams of significant safeguards concern are highly heterogeneous and otherwise
intractable. This situation has led to the development
of nondestructive methods, which apply state-of-the-art
knowledge of nuclear physics and instrumentation to the
problem of measuring nuclear material in a rapid and
convenient manner.

Nondestructive analysis lends itself to nuclear material
measurement since (1) it does not impair future usefulness
of material and (2) it can measure highly radioactive material
on site with a minimum amount of handling and sample preparation. These methods may be classified as passive techniques
which utilize the intrinsic radiation of nuclear materials,
and active techniques which depend on properties induced
by various types of external irradiation. In addition,
the heat associated with the alpha emission of Pu isotopes,
notably ^{238}Pu, has given rise to the rapidly developing
and highly accurate technique of calorimetry. A summary
of techniques is shown in Table I, and typical materials
amenable to NDA techniques are listed in Table II. Both
active and passive non-destructive techniques are employed
at DOE facilities.

The precision attainable by NDA depends on both the
technique used and the stream measured, and can range from
about 20% to about 1%. In many cases large uncertainties
are acceptable, either because the stream involved does not
make a major contribution to the measured material balance
or because the material also lends itself to accurate
chemical methods and NDA is invoked mainly as a rapid,
semi-quantitative confirmatory measurement (on the part of
either the facility operator or an inspector). A source
book for reliability data relevant to NDA measurements was
published by LASL in 1977 (4).

NDA technology sponsored by DOE includes development
and application of NDA methods to all phases of the fuel
cycle, inspection and assay of materials. In addition,
methodologies that are fully developed, partly developed,
or in a design stage show much promise for materials management, quality and process control, safeguards inspection,
and criticality safety. The application of NDA to all
aspects of inspection and quantitative assay of special
nuclear materials (SNM) will provide a rapid and accurate
determination of material balances to detect unauthorized
losses or diversion of SNM.

U. S. research and development activities in NDA
are conducted principally at the Los Alamos Scientific

Table I

Major Non-Destructive Analysis Methods

1. Active neutron assay

    Principle: Irradiation of sample with neutrons from
    14MeV neutron generators, Van de Graaf accelerators
    or radioactive sources (^{252}Cf, Sb-Be, ^{238}Pu-Li),
    and observation of the resulting fission neutrons
    and/or gamma rays.

2. Passive gamma assay

    Principle: Detection of gamma rays entitled by fissile
    isotopes (mainly ^{235}U and ^{239}Pu), using NaI and Ge(Li)
    detectors.

3. Passive Neutron assay

    Principle: Neutron coincidence counting of ^{240}Pu
    spontaneous fission, using high-efficiency detection
    systems.

Table II

Typical Streams Amenable to NDA Techniques

1. Feed materials

   a. UF_6 of various enrichments in large cylinders

   b. Plutonium nitrate solutions in 10-liter bottles

2. Intermediate materials

   a. Blended powder

   b. Pellets at various stages of fabrication

   c. Uncoverable scrap

3. Product Materials

   a. Pellets

   b. Fuel plates

   c. Fuel rods

   d. Fuel bundles

4. Dilute solid waste

Laboratory (LASL), Lawrence Livermore Laboratory (LLL), and
Mound Laboratory.

   Developments in NDA have proceeded along several lines.
At one extreme, the techniques developed center on hand-held,
portable devices which can be used both by plant personnel
and by inspectors. At the other extreme there are instrumented vans for use, for example, by NRC inspectors (developed at BNL and ORNL), and sophisticated mobile laboratories
for active interrogation, e.g., the GAMAS and MONAL system
developed, respectively, by Gulf Radiation Technology (now
IRT Corporation) and Los Alamos Scientific Laboratory. A
variety of devices occupy a middle ground between these
extremes, as illustrated in Table III. Devices of this kind
are today in common use, both in private industry and Government-
owned facilities, at home as well as abroad. They make an
essential contribution to the task of establishing material
balances based wholly on measured values. A few more
specific aspects of NDA are discussed elsewhere in this
Symposium.

   Work is underway to integrate the newly developed measurement techniques into a balanced, cost-effective safeguards
and security system based on the concept of defense-in-depth.
Many of the processes envisioned for the new fuel cycles
have typical counterpart processes in DOE facilities. A
project is underway to demonstrate dynamic materials control
(DYMAC) by instrumenting the LASL plutonium recovery/fabri-
cation operations, and demonstration of automated inventory
techniques. In essence, the DYMAC system incorporates the
following techniques: (1) an in-line measurement system,
employing NDA instrumentation, to give real-time assay data
from predetermined measurement points; (2) a direct and
automated transfer of data from the facility area into a
central computer which displays the data at determined
measurement posts; (3) the use of an automated accountancy
system for the purpose of providing a rapid status on material balance for smaller, controllable units of a plant.

   Both chemical and NDA methods must rely for their
success on the availability of suitable standards which are
traceable to a national system of primary standard reference materials. The National Bureau of Standards (NBS) has
for some years distributed a limited number of such materials.
With an expanding nuclear industry it has become apparent
that additional primary standard reference materials must
be made available, both to facilitate traceability and to
provide standards representative of the increasing number
of materials which the measurement system must address. As
a result of studies undertaken by NRC, DOE and NBS, a
National Nuclear Standards and Measurement Assurance Program
has been designed to cover the whole area of standards and
measurement control, field testing, performance evaluation

Table III

Typical Established NDA Devices

NDA Type	Device	Type of Nuclear Material Measured	Developers	Systematic (1) Error (2σ)
Interrogation by neutrons (active neutron assay)	Scanner for low enriched fuel rods (Cf252 source)	U^{235} in LWR rods	LASL	0.5%
	"Random Driver" (Am-Li source)	U^{235} in high enriched feed and scrap	LASL	
	Isotopic Source Assay System (ISAS) active mode (Cf252 source)	$U235$ in 18% enriched scrap, fuel pellets, rods and plates; $Pu239$ in waste	IRT (Intel-com-Rad. Tech.)	6%
Passive gamma-ray assay	NaI and GeLi fuel rod scanners	Pu in mixed oxide fuel rods (2-25% Pu)	General Electric	1%
	Enrichment(2) meters	$U235$ in UF_6, low enriched fuel rods, pellets, incinerator oxides, etc.	LASL	10%
	L-10 bottle assay unit	Pu in $Pu(NO_3)_4$ receipts	Idaho National Engineering Lab.	10%

Table III (cont)

NDA Type	Device	Type of Nuclear Material Measured	Developers	Systematic Error (2σ)
	GeLi segmented gamma scanner	U^{235} in high enriched scrap and waste (up to 30 gal. drums)	LASL	10%
		Pu in various materials		10%
	GeLi low level package assay unit	Pu in solid glove box waste	Numec	20%
	Low level waste gamma scanner (GeLi or NaI)	Pu in low level waste, up to 55 gal. drums	Westinghouse	40% (Pu not separated) 10–12% (Pu separated)
	Eight channel drum scanner	Pu in scrap and waste (55 gal. drums)	LASL	20%
	NaI barrel	Pu and U^{235} in waste	IRT	15–30% for Pu, depending on standards available 20–60% for U^{235} depending on nature of waste
Passive neutron assay (neutrons emitted from spontaneous fission events)	Isotopic Source Assay System (ISAS) (passive mode)	Pu in waste (1 gal. containers)	IRT	10%

Table III (cont)

NDA Type	Device	Type of Nuclear Material Measured	Developers	Systematic Error (2σ)
	Neutron coincidence counters	Pu in feed, scrap and waste material (contained in cans up to 1 gal., bags, etc.)	Brookhaven National Laboratory, LASL	10%
Calorimetry		Pu in a wide variety of matrices	Mound, Argonne National Laboratory	0.1%

(1) Random errors can be reduced to substantially below the systematic errors.

(2) Because of the pronounced attenuation produced in the dense samples involved, the 185 Kev gamma activity obtained from U^{235} is proportional to the enrichment.

and assay system optimization. The program is being coordinated with NRC, industry, DOE and its laboratories. Significant aspects will be discussed in this Symposium by participants from these organizations. The program will also be
coordinated with various sample exchange programs which
have played an important role in safeguards measurement
for some years. The best known of these is the Safeguards
Analytical Laboratory Evaluation (SALE) Program, adminis-
tered by NBL. The goal of this voluntary program is to
provide a means, through periodic interlaboratory comparison,
by which laboratories can demonstrate continued proficiency
in safeguards measurement of nuclear material, and to
provide evaluation and assistance to those laboratories
which are not measuring to the degree permitted by the
methods which they choose to use.

     The NBS function in the national safeguards system is
to provide primary national standards reference material to
the nuclear community for chemical and isotopic analysis
of plutonium and uranium and for radioactivity measurements.
In addition, NBS lends its expertise to other safeguards
activities. These include calibration services related
to mass, volume, voltage, temperature and radioactivity.

     U. S. activities in the measurement quality assurance
area are being coordinated with similar activities abroad,
especially at a variety of IAEA-affiliated laboratories
and at various regional laboratories.

SUMMARY AND THE MEASUREMENT CHALLENGE

     In summary, "Safeguards Needs in the Measurements Area"
and the associated realm of measurements have been described.
The increasingly stringent safeguards requirements, necessi-
tated by a growing nuclear industry, have placed severe
and growing demands on SNM measurement capabilities. To
meet these unprecedented requirements, it has been essential
for DOE to develop new measurement technology, incorporated
with appropriate physical and security procedures, for safeguarding nuclear materials.

     The new technologies are being passed on to the nuclear
community. Acceptance of these technologies is achieved
through evaluations and participation by the nuclear community in various measurements and standards groups and the
application and standardization of measurement control
techniques. As a result, we are not only keeping abreast
with present nuclear safeguards technology, but the programs
are also consistent with the projected growth of the U. S.
nuclear industry and the needs of international safeguards.

     Recent technology and hardware development provide
the techniques and equipment necessary to implement essentially hands-off "short-term" internal control systems,

periodic "longer-term" assessments, and where necessary
limited access to sensitive nuclear technology such as for
uranium enrichment. In-plant nondestructive assay instrumentation is being coupled with automated data processing
equipment to provide essentially continuous accounting
and control of nuclear materials on a detailed unit process
basis consistent with today's safeguards requirements as
well as plant requirements for process and quality control,
criticality control and health and safety.

If on the other hand, there are undetected losses,
undetermined materials present, or poor measurements of
material on hand, responses to these requirements suffer
and the opportunities for diversion or theft are much
greater.

With the increased availability of attractive nuclear
materials, the risks are compounded; thus safeguards practices must be continuously updated to keep in phase with the
threats and opportunities to thwart the peaceful use of
nuclear power.

To assure that risks to society are acceptably low,
the safeguarding of nuclear material is a vital and challenging task -- one that requires an accurate measurement
system and high quality assurance against malevolent practices and the stated needs of cost-effective in-depth
safeguards.

While "conventional" material balance accounting
functions are necessary for a totally integrated system,
recent efforts by LASL (5) and others in developing safeguards designs for generic facilities have identified
inherent limitations in sensitivity and timeliness. For
example, measurement uncertainties desensitize the system
to certain losses of SNM for high-throughput facilities.
Timeliness of the system is limited by the frequency of
physical inventories and practical limits on how often a
facility can shut down or how well an inventory can be
performed under continuous or semi-continuous operations.

One of the most challenging analytical problems in the
nuclear fuel cycle concerns the measurement of irradiated
input solutions to reprocessing plants. This problem
has been attacked in a number of ways, most notably by
isotope dilution-mass spectrometry which yields both chemical concentration and isotopic ratios of plutonium and
uranium, and direct x-ray fluorescence or densitometry in
solution which yields the chemical concentrations of both
uranium and plutonium especially important in coprocessing
modes of alternative fuel cycles. DOE is testing systems
of this type at the Savannah River Plant, with a view to
developing on-line methods for analyzing input solutions.
A related method has been automated at the Kernforschungszentrum, Karlsruhe, W. Germany. Many chemical and NDA

techniques in existence require macro amounts (milligram to gram amounts or greater) of nuclear material for elemental analysis of plutonium, uranium and their related isotopes. The transportation and handling of macro quantities of nuclear materials and related shielding in and out of the laboratory is relatively expensive, time-consuming and cumbersome. Presently the shipment of plutonium analytical samples via aircraft is not permitted.

DOE is sponsoring a task at the Oak Ridge National Laboratory to develop a resin-bead collection technique for applicability, in particular, to spent fuel material which would require one microgram or less of Pu or U on the bead for analysis. This would reduce the amount of Pu in a given sample to about 3×10^{-9} curies.

State-of-the-art conventional measurement methods, recently developed measurement technology, special in-plant sensors, plant instrumentation signals, and data-analysis techniques supported by computer and data-base management technology are being used and further innovations are being sought to meet the challenges for rapid, accurate measurements, control and verification of nuclear materials in the various physical and chemical forms throughout the nuclear fuel cycle.

ABSTRACT

An effective safeguards measurement system must cover a multitude of material forms ranging from essentially pure substances to highly heterogeneous materials. In addition there are varied and sometimes conflicting demands for accuracy and timeliness. Consequently, a judicious and systematic choice must be made between methods based on sampling followed by chemical analysis or nondestructive methods based on nuclear properties. Fundamental advances in analytical chemistry made during the years preceding World War II enabled Manhattan Project scientists to develop methods which contributed to the success of both the immediate goal and the developments which have taken place since. Examples will be given of evolutionary developments in the direction of timeliness through varying degrees of automation. Nondestructive methods, first introduced because of the need to measure scrap and other intractable material, are finding broader areas of application.

Aided by DOE-sponsored research and development, new techniques providing greater accuracy, versatility and timeliness are being introduced. It is now recognized that an effective safeguards measurement system must make concerted use of both chemical and nondestructive methods. Recent

studies have fostered understanding of the relative importance of various process streams in the material balance
equations and have highlighted the need for a systematic
approach to measurement solutions for safeguarding nuclear
materials.

LITERATURE CITED

1. Jaech, John L. "Statistical Methods in Nuclear Material
    Control" Technical Information Center, Office of Informa-
    tion Services, USAEC, 1973.

2. Jones, Ralph J., compiler and editor, "Selected Measure-
    ment Methods for Plutonium and Uranium in the Nuclear
    Fuel Cycle" TID-7029 Division of Technical Information,
    USAEC, 1963.

3. Rodden, Clement J., editor, "Selected Measurement
    Methods for Plutonium and Uranium in the Nuclear Fuel
    Cycle," second edition, Office of Information Services,
    USAEC, 1972.

4. Reilly, T. D., Evans, M. L., "Measurement Reliability
    for Nuclear Material Assay," LA-6574 Los Alamos
    Scientific Laboratory, 1977.

5. Shipley, J. P., Cobb, D. C., Dietz, R. J., Evans, M. L.,
    Schelonka, E. P., Smith, D. B., Walton, R. B., "Coordi-
    nated Safeguards for Materials Management in a Mixed-
    Oxide Fuel Facility," LA-6536, Los Alamos Scientific
    Laboratory, 1977.

RECEIVED JUNE 12, 1978.

2

Standards for Chemical or NDA Measurements for Nuclear Safeguards—a Review

CARLETON D. BINGHAM

U.S. Department of Energy, New Brunswick Laboratory, Argonne, IL 60439

The objective of nuclear materials safeguards
is the prevention of successful malevolent acts involving
nuclear materials and facilities. Safeguards consist of an
integration of physical protective measures, materials
control, and materials accountability. This symposium
fittingly addresses the subject of measurements of nuclear
materials for safeguards purposes.

Accuracy of measurements is essential to define
the quantity of material which is on hand and which must be
safeguarded. A facility inventory statement, based on
accurately measured values, which agrees with the book
inventory is one means by which the absence of diversion is
demonstrated. The precision with which measurements are
performed becomes the basis for the uncertainty attached to
an inventory statement. When measurements can be performed
to a higher degree of precision, the overall uncertainty is
reduced and the sensitivity for detecting a diversion is
enhanced. In order to make authoritative statements about
measurement accuracy and precision, a measurement assurance
program is necessary. Within the framework of a measurement
assurance program, it is relatively simple to demonstrate
precision and make statements regarding a random error of
measurement. To demonstrate accuracy and make statements
regarding bias and/or systemmatic error require that
measurements be traceable to accepted reference bases.

Safeguards have placed an increased emphasis
on nuclear materials measurement assurance during the past
decade. This emphasis exists both in the U.S.A. and
throughout the international nuclear community. In the
U.S.A., measurements on a complex variety of materials from
all parts of nuclear fuel cycles are required to be
traceable to a national measurement system.(1) Implicit in
such a requirement is the existence and availability of
appropriate means for demonstrating such traceability. Let

us examine this implication in light of a hypothetical
safeguards measurement. A batch of material of given (i.e.,
observed) mass is to be measured to confirm an addition to
an inventory stratum. The bulk material is sampled
according to a statistically designed plan giving due
concern to gain or loss of moisture, oxidation, etc., and
the mass of the individual samples is determined. Each of
the samples is dissolved following an accepted procedure and
the mass (or volume) of the resultant solution is
determined. From each of the solutions, aliquants (either
by weight or volume) are taken upon which the measurements
will be performed. I will address measurements later in the
paper. (At this point, even before chemical assay
measurements have been performed, measurement assurance
requires calibrated balances, thermometers, and barometers,
the latter for buoyancy corrections, to obtain accurate
masses or similar calibrations where volumes are concerned.
These calibrations for mass or volume are every bit as
important to the final quality of measurement as the
measurement itself, but are often taken for granted. Aside
from mentioning these measurements being an integral part of
safeguards measurements, I shall not dwell further on them.)
    The U.S. National Bureau of Standards (NBS)
is charged with the responsibility for establishing and
maintaining a national measurement system. In addition to
the basic units: mass, length, time, etc., the Bureau
establishes measurement technology which is disseminated to
users.
    Reference materials are one means for
transferring measurement technology to users inasmuch as
instruments and chemical reagents can be calibrated to
reproduce the value assigned to the reference. Reference
materials form the base for defining the quality of a
measurement method - its accuracy, precision, and
"ruggedness". These materials are also the means of
relating measurements made at different sites to each other.
A hierarchical structure of reference materials exists in
measurement science such that lower levels are derived from
upper levels.
    Primary standards are high-purity elements or
compounds, the validity of whose reference values (assigned
according to the best available scientific procedures) can
be assured when the material is treated according to
instructions on the certificate. This implies a certain
known chemical stability or an ability to achieve and
reproduce a known compositional stoichiometry. Standard
reference materials (SRM's)issued by NBS constitute the
primary standards for the chemical or isotopic measurement
of uranium and plutonium and provide the link of
traceability to the national system. Some of the nuclear

SRM's are related to other primary chemical
standards such as the redox standards $K_2Cr_2O_7$, As_2O_3, and
$Na_2C_2O_4$.

Primary standards represent a valued resource
and should be used when there are no other suitable
alternatives to demonstrating the traceability link. The
preparation and characterization of primary standards is
expensive in money and time and requires a highly technical
effort by experienced scientists. Primary standards should
not be used routinely to prepare working solution standards
or bench control standards.

Secondary standards are prepared as elements
or compounds of varying levels of purity. Their chemical
stability may be such that large batches cannot be prepared
to provide long-term supplies. Their stability, on the
other hand, may be similar to that of their primary
counterparts and only the purity and uncertainty in their
assigned values may be different.

Secondary standards, such as those distributed
by the Department of Energy's New Brunswick Laboratory (NBL),
provide an alternate path to traceability. These standards
represent typical materials found in current nuclear fuel
cycle technology. Materials available cover the range from
ores and counting standards, enrichment plant product,
conversion plant intermediates and product, and production
plant product. These materials may be more representative
of those in actual plant or facility use and, as such, may
offer a more simple, but not necessarily as certain, path of
traceability.

To what extent is the implied availability of
reference materials being met? A group, convened in
November, 1977, to advise the International Atomic Energy
Agency (IAEA) of the current status of chemical and isotopic
reference materials in the nuclear fuel cycle, made the
observation that the primary nuclear reference materials
generally available to the nuclear community provide the
means to demonstrate the traceability of safeguards
measurements. The group recommended that effort be expended
to provide a greater variety of secondary reference
materials and to improve the quality of some existing
primary reference materials.(2)

Where suitable reference materials are not
available from external sources, a facility must resort to
working standards prepared internally. These materials
generally are prepared from actual material streams in the
fuel cycle. This path to traceability is not always
straightforward - one may need to exercise considerable
ingenuity to assure that all steps can be traced to the
national system - but traceability can be achieved.

It is essential that the measurement method

used in a given facility undergo thorough and complete
testing by the developer or the user to document the effect,
if any, of impurities, cationic and anionic, and the effect
of slight unintentional variations to the published method
(i.e., how "rugged" is the method). Here is where reference
materials come into play. Reference materials are essential
for the initial statistical testing of a method - i.e.,
establishing initial estimates of accuracy and precision.
The SRM should then be used to compare the response of the
method to a secondary or working material of lesser purity.
When comparability of observations can be demonstrated, then
the less expensive secondary or working material can be
selectively doped with impurities to document the effect of
impurity levels and to ascribe accuracy and precision when
typical materials are measured. The "ruggedness" should be
studied with these latter materials.

A commonly heard request is for a reference
material of a given compound to match the compound of
interest at a given site. An inability to trace the
measurement through all of the analytical steps is cited as
basis for the need. The stability of the material requested
notwithstanding, the chemistry of the measurement process
needs to be examined. Where uranium matrices, for example,
are concerned, once solubilization is effected, for all but
unusual cases, converting the solution to a nitrate form,
presents the uranium for measurement in the same chemical
state in solution as is achievable with existing metal or
oxide primary or secondary standards. Now the possible
effect of impurities needs to be considered. If the
measurement method being used exhibits a response which is
affected by the presence of impurities known to be present,
the solutions of standards may require doping to an
appropriate cationic or anionic impurity level. A more
satisfactory option would be to change to a method which is
not affected by the impurities known to be present. It
suffices next to demonstrate that total solubilization has
occurred. Filtration of the solution, followed by visual or
radiometric observation for a residue, is usually an
adequate demonstration. If a residue is present, technology
exists for solubilizing it, i.e., fusions, combustion
followed by acid treatment, etc.(3)

My discussion thus far has been limited to the
traceability of destructive chemical analysis. The potential
advantages inherent in nondestructive assay (NDA) have
attracted much attention both at the R&D level and at the
user level. These advantages equate to near real-time
measurements of process streams and the ability to measure
heterogeneous composites in the scrap/waste stream where
sampling errors override any accuracy and precision
advantages of destructive chemical analysis. In the NDA

area, traceability becomes more difficult, but not
impossible as some maintain.(4,5) Practically all NDA
measurements exhibit some form of matrix dependence which
prevents their being an "absolute" measurement and thus
requires that the instrument or system be calibrated by
reference to known quantities of material. This known
quantity of material must be contained in a matrix which has
chemical and physical properties similar to the process
materials to be measured before the calibration can be
transferred to those materials. Knowledge of the effect of
differing chemical and/or physical properties on the
measuring system can be used to apply an empirical factor to
the observed system response to obtain an estimate of the
container contents.

In the fuel cycles with which there is considerable
experience, viz, LWR, LMFBR, and HTGR, there are site-
specific differences in fabrication practices, e.g.,
burnable poison vs no burnable poison, coprecipitation vs
mechanical blending, resin-bead-converted low-density
carbide vs high-density mixed carbide, etc., which yield a
product with differing physical properties and thus preclude
a centralized national source of a practically managed
finite number of NDA reference standards. Granted, there
are no off-the-shelf primary or secondary standards from a
nationally accepted source; however, via the primary and/or
secondary materials that do exist nationally, a facility can
synthesize working standards or analyze representative
samples of a material preparation in a manner that
traceability can be demonstrated. With the current state-
of-the-art of NDA measurement technology, which is highly
matrix dependent or material specific, traceability of NDA
measurements must be through working standards which have
been prepared and/or characterized by measurement technology
traceable to the national system.

NBS is collaborating with the European Economic
Community (EEC) in the characterization of a joint US-EEC
reference for the NDA of product low-enriched uranium oxide.
This may enable more timely measurements of product to be
performed to increase the confidence in the inventory
statement.

NBL is preparing two prototype NDA secondary
standards containing enriched uranium based upon the
response by NDA users to a questionnaire attempting to
define common areas of application and need. These are
expected to be available for evaluation by comparative
measurements in late 1978.

To what extent does the quality of existing
reference materials meet the needs of the nuclear community,
where are improvements required, and how are these
requirements implemented? In the IAEA Advisory Group

meeting mentioned earlier, it was acknowledged that there were several areas in which improvements to the quality of existing reference materials should be made. Existing plutonium isotopic standards were originally measured and certified using mass discrimination data obtained using separated uranium isotopes, not plutonium. Plans already exist to remeasure plutonium isotope ratios and correct the observed ratios on the basis of mass discrimination data from separated plutonium isotopes. Certain uranium isotopic standards should also be certified for uranium content so that systemmatic errors are not introduced into isotope dilution measurements when a "U_3O_8" stoichiometry is assumed. Certain uranium and plutonium assay standards should also be certified for isotope abundance distribution thereby allowing one standard to suffice where two are now required. More accurate plutonium and americium half-lives are necessary in order to transform calorimetric measurements into measurements of a quantity of plutonium. NBS and other national standards organizations are moving to implement these recommendations.

As the state of routine measurement art improves, there needs to be a corresponding reduction in the uncertainty assigned to reference materials used to calibrate such measurements. This further requires the development of improved methods for characterizing primary reference materials. A reference material certified to 0.1% is of little use to make accuracy statements for measurement methods routinely capable of 0.05% RSD precision. Materials which are now certified to $\pm0.02\%$ of the assigned value may require in the near future a recertificiation to 0.002% so that the uncertainties in secondary or working standards are less subject to the effect of the uncertainty in the reference value of the primary standard. This additional decade of certainty will be expensive to provide, but if attained, can represent a strengthening of nuclear materials safeguards in that reduced precision and increased accuracy translate into a greater sensitivity for detecting a diversion which, in part, meets the objective of safeguards.

Literature Cited
1. Code of Federal Regulations, Title 10, Part 70, Section 70.57.
2. Report of the Advisory Group to the International Atomic Energy Agency on Chemical and Isotopic Reference Materials in the Nuclear Fuel Cycle, held November 8-10, 1977, Vienna, Austria. To be published.
3. Bingham, C. D., Scarborough, J. M., and Pietri, C. E., "Safeguarding Nuclear Material", Proceedings of Symposium, Volume II, pp 107-115, International Atomic Energy Agency, Vienna, 1976.

4. Bishop, D. M., Nuclear Materials Management (1976),
    Volume V (No. 1), pp 16-27.
5. Bingham, C. D., Yolken, H. T., and Reed, W. P.,
    Nuclear Materials Management (1976), Volume V (No. 2),
    pp 32-35.

RECEIVED MAY 1, 1978.

Nuclear Safeguards and the NBS Standard Reference Material's Program

W. P. REED and H. T. YOLKEN

National Bureau of Standards, Washington, DC 20234

The commercial use of fissionable material as an energy source is predicated on the assumption that the general population will not be exposed to undue risks. One of the risks considered is the possible diversion of fissionable material for the clandestine manufacture of nuclear explosives. Reduction in risk from this sort of activity is one of the purposes of Nuclear Safeguards. As such, "safeguards" are composed of various elements including physical protection, materials control, and accounting procedures to determine the amount and location of nuclear materials. In this discussion we wish to limit ourselves to the measurement of these quantities in the accounting type procedures used for safeguards.

Any accounting procedure will only work as long as the units of measure are equivalent. Thus, when one does financial accounting one always talks in terms of a single reference -- the most common reference in the United States being the dollar. Other references are possible, of course. Such possibilities as the deutschmark, pound sterling, or yen, come to mind. But, within any single accounting system, the amount of currency going out, amount of currency being received, and amount on hand, must balance. The system will not balance if different currencies are used unless the currencies are converted to a base currency by means of compatible exchange rates. Thus, the need for a single base currency as dollars.

This simple analogy, also applies for safeguards accountability systems. In order to work they also have a base currency, i.e. one set of base standards. This need for a single set of base standards results in requirements that stipulate that all measurement be made with reference to the National Measurement System or that all measurement be "traceable to NBS". NBS then, in turn, must assure compatibility with internationally recognized standards.

In an attempt to provide a currency for safeguards accountability measurements, an empirical but compatible standardized set of references could work and in some areas of the physical sciences this is exactly what is done. However, the drawback of this system is that the properties being measured relate back only to the properties of the standardized materials and without these standardized materials, the properties being measured are not relevant.

If however, the currency for safeguards accountability measurement is based on accuracy rather than a series of empirical reference materials, then the currency of the accounting will be the actual number of atoms of uranium or plutonium present in the measurement, rather than a numerical value related to a series of standards.

The net result of an accuracy based system is that anyone who makes measurements - taking care to evaluate systematic bias as well as random error - can, with some confidence, be sure he is, within the limits of his uncertainty, traceable to the National Measurement System. The demonstration of that traceability for regulatory purposes is another question, however, and should be answered by the regulatory agency.

It becomes apparent that one of the goals in certifying Standard Reference Materials for safeguards purposes (as well as most other purposes) is to provide accurate certification values and to evaluate the systematic and random errors associated with these values. If this is done, then Standard Reference Materials will provide a convenient meter stick with which to demonstrate one's analytical capability and reference to the National Measurement System. But it also is important to note that these SRMs are not imperative to making the required measurements.

Although standards, or SRMs, are not imperative to achieve traceability or to make good measurements, they do make the task of achieving accurate measurement much easier and to that extent it is appropriate to ask what NBS SRMs are available for the measurement of safeguarded material. Table I illustrates these materials. While the list is not extensive it does include all measurements necessary to provide the accurate analysis of about all of the material currently considered for safeguarding. Note that for the uranium fuel cycle these are both metal and oxide materials for uranium assay purposes as well as a complete line of isotopic standards for the calibration of isotopic measurements. In addition, for plutonium measurements, plutonium metal and plutonium sulfate are available for assay purposes and plutonium sulfate isotopic standards are available for isotopic measurements.

With these reference materials, analytical laboratories can make the measurements necessary for determining the basic safeguards quantities, i.e. atoms of uranium and the atom ratio of the isotopes and atoms of plutonium and the atom ratio of the isotopes. These standards also allow measurement laboratories to assess the uncertainty of their measurement.

It is obvious to all who make measurements for safeguards purposes that while the basic standards are available, many useful and almost necessary reference materials are not available. Needed standards include materials similar to typical nuclear fuels,

Table I

Currently Available Special Nuclear
Standard Reference Materials

SRM Number	Description	Certification
950b	Uranium Oxide (U_3O_8)	Uranium Content
960	Uranium Metal	Uranium Content
949e	Plutonium Metal	Plutonium Content
944	Plutonium Sulfate Tetrahydrate	Plutonium Content
945	Plutonium Metal	Trace Elements
U-0002 thru U-970 (18)	Uranium Oxide (U_3O_8)	Isotopic Abundance
946	Plutonium Sulfate Tetrahydrate 76% to 92% Pu-239	Isotopic Abundance
947		Isotopic Abundance
948		Isotopic Abundance
993	Uranium Solution (U-235 SPIKE)	Uranium Content and Isotopic Abundance

specifically oxide materials as UO_2 and PuO_2 and mixed oxides, NDA standards for scrap and waste measurement and even ore and UF_6 standards for the processing of nuclear material. Newly developed or alternate fuel cycles will substantially enlarge the list.

Which of the above needs are of highest priority and how to go about meeting these needs are the questions that are raised. The questions are raised of necessity since the resources at NBS are not sufficient to address all of these problems at once and even if possible, it would be unwise to do so.

First, it should be noted that reference materials are available from other sources and these should not be duplicated, if resources are allocated wisely. One good source of reference materials is the New Brunswick Laboratory. NBL has a well recognized and respected standards program that relies in part on the traceability or reference of their standards to NBS standards. In addition, NBS and NBL carry out sample interchange and cooperative certification efforts to assure the compatibility of NBS and NBL standards. NBL is thus in the best position to certify many of the necessary secondary standards needed in the nuclear industry since this work is basic to their primary mission.

A second consideration concerns those materials that in most cases should not be used as reference materials. Some materials by their very nature are not sufficiently stable to be used as SRMs unless no other alternative exists. For example, if the base weight of a material cannot be reproduced in all laboratories in which it is used with a reasonable uncertainty, it will not make a useful assay standard. The simple logic of the situation should be apparent. That is, the uncertainty of the base weight of the material must be included and added to the uncertainty of the measurement in order to provide a certifiable value. Thus, for example, a uranium value measured to $\pm$ 0.02% on UO_2 may have to be either certified at $\pm$ 5% to include the laboratory-to-laboratory variability due to the differing conditions under which the sample would be weighed before measurement or have overly rigid constraints placed on how it is to be handled in the laboratory.

This sort of problem precludes the issuance of several materials as primary SRMs and makes their use for other kinds of standards subject to great caution. UO_2, PuO_2, and mixed oxides for most purposes fall in this category and for this reason it is far better to reference the measurement of these materials to the measurement of stable reference materials wherever possible. The scientific evidence demonstrating this reference and verifying the accuracy of the measurement becomes part of the chain of traceability and is just as valid (even if more cumbersome) than the use of primary reference materials of the same matrix.

The word secondary may be used for several different purposes. In this case its use is to indicate the relationship of these standards with primary standards.

This now leads to the current program expansion for the certification of Standard Reference Materials at NBS for the nuclear
community. This expansion is a consequence of the recently started major new NBS effort in measurements and standards for nuclear
safeguards. This program is focused on providing Reference
Materials, calibration methodology, calibration services, reference
methods of measurement, and data needed for all types of measurements for nuclear safeguards. These include bulk measurements,
chemical and isotopic analysis, passive and active nondestructive
assay, and associated statistics and sampling methods. The program
is focused on providing standard services to the nuclear community
and is supported by NRC and DOE.

Turning now to new SRM work, the highest priority is the
placing of the plutonium isotopic SRMs on an absolute basis. As
mentioned earlier, the basic standards for the measurement of the
number of atoms of each isotope of plutonium are available. However,
because of health physics requirements, the isotopic measurement
work at NBS was terminated several years ago. Consequently, measurements necessary to ascertain the magnitude of the correction for
bias in the thermal filiment ionization process for the plutonium
isotopes has not been performed. The current certified data are all
based on the assumption that the filiment bias correction is similar
to that of uranium. This is probably a close approximation but not
exact. With the advent of the measurements for Nuclear Safeguards
Program at NBS, funds have been provided by DOE for a new plutonium
laboratory. A contract has been let for this laboratory which will
be located within the NBL facility at Argonne, Ill. When these
facilities are completed (1979) this work will be undertaken.

While NBS has available a series of isotopic standards for the
isotopic measurement of uranium, there is still a need for relating
these standards to isotopic measurement of UF_6. This need is a result
of the high precision with which gas mass spectroscopy measurements
are made and the inadequacy of referring these measurements to the
NBS solid U_3O_8 standards. This project will consist of setting
aside large quantities of UF_6, depleted, normal and enriched UF_6,
and making a series of intercomparisons between gas and thermal-
ionization mass-spectrometric measurements of these materials. This
material will be made available to gas mass spectroscopist in
sufficient quantity that they can compare the certified standards with
their own standards and prepare blends for those ratios which fall
in between those of the certified standards.

The need for highly precise and accurate measurements usually
leads to the discussion of the use of the isotope dilution mass
spectrometric measurement technique. This is a technique employed
quite successfully in the analysis of trace quantities of material
and has many inherent advantages because the chemistry need not be
quantitative and the errors and uncertainty in the measurement are
usually those associated with the instruments. Consequently, the
precision and the bias of the method can more easily be evaluated.
In this area NBS has already issued SRM 993 which is a U-235 spike
material. This material was intended for a variety of uses including

geological measurements and nuclear inventory measurement. However,
because the U-235 content of many items in the nuclear materials
inventory is high, the use of this SRM as a spike is awkward. For
this reasons, NBS is starting to prepare a Uranium-233 SRM for use
as a spike. This material should be of greater utility in the
nuclear industry.

In a similar manner, NBS is working with DOE, Los Alamos
Scientific Laboratory for the preparation and certification of a
Plutonium-244 spike solution for the measurement of plutonium and
plutonium bearing materials. This work has been hampered in the
past due to the lack of pure Plutonium-244 material. However,
sufficient material has been made available to make it possible
to prepare a lot of SRMs (approximately 200 spikes) containing 5 mg
of plutonium-244.

The use of Non-Destructive Analysis (NDA) to verify the source
and fissile content of production materials, scrap and waste is
a major part of the safeguards program for most facilities.
Unfortunately, the verification of the accuracy of the measurement
and the uncertainty estimate are most difficult. The need for
reference materials and document standards in this area is vital.
Unfortunately, this need is compounded by the variety of measurement
systems used and the constraints put upon the container size and
content by these systems. The current trend to have the various
facilities prepare their own reference materials verified by
chemical assay or independent measurement of production items is
likely to continue. This is so, especially, since it is difficult
to conceive of one laboratory capable of preparing all of the sizes
and types of standard needed.

Part of the nuclear safeguard effort at NBS has been devoted
to developing capabilities in the NDA area with the intention of
offering a relatively small number of primary NDA SRMs. These
materials will be designed to be useful in a wide variety of fuel
cycle facilities. Examples include U_3O_8, uranium metal, plutonium
metal, and plutonium oxide (all at varying isotopic compositions).
The facilities independent SRMs would be in contrast to facilities
specific or unique SRMs for scrap, fuel rods, waste, etc. This
approach is consistent with a recent recommendation of the
International Atomic Energy Advisory Committee on Physical Standards
for NDA. They also recommended that facilities and inspectors
jointly develop facilities dependent or unique reference materials.
The contents of these reference materials would be verified by NDA
measurements and in many cases by destructive analysis of samples
taken before fabrication.

Work at NBS is currently going on in the areas of passive
gamma, neutron interrogation and calorimetry measurement. At this
time, arrangements have been made for the preparation and certification of a series of slightly enriched Uranium SRMs for passive
gamma measurement of the uranium-235 and uranium-238 content. These
SRMs will be used to calibrate procedures for the measurement
of U_3O_8, UF_6 and UO_2. This work is part of a cooperative program

with EURATOM with costs and measurements being shared. The material will be prepared in Europe, characterized by the EURATOM Laboratory in Geel Belgium, NBS and NBL, packaged at Geel and the final measurement made at various EURATOM Laboratories, NBS, NBL and LASL. Final data evaluation, and certification will be performed at NBS for all standards issued in the United States. The standards will consist of about 200 grams of U_3O_8 of 1, 2, 3, 4 and 5% enriched Uranium-235 sealed in metal containers. While this will not answer the need of many NDA users it should serve as a starting point among NDA users and standards laboratories. In addition, International Atomic Energy Agency inspectors will observe the fabrication of these SRMs and validate them for use by the IAEA. These U_3O_8 SRMs will, in effect, be the first international NDA reference materials. We expect upon completion of the low enriched U_3O_8 SRMs to expand the series (hopefully again in a cooperative mode) to include high fissile content U_3O_8.

In another area, that of plutonium assay by calorimetry measurement, NBS has participated in the measurement of several of Mound Laboratories sealed plutonium sources. NBS used an ice calorimeter as a reference method to check the results of Mound's ice calorimetry results. Agreement was good. The NBS calorimetry group has made arrangements to borrow a heat flow calorimeter from Mound specifically suitable for the measurement of plutonium heat sources. When this equipment is set up, calibrated, and compared to the ice calorimeter, NBS hopes to be in a position to provide a mechanism for demonstrating traceability to those who use this NDA technique. The exact form of this mechanism is now unclear and while the issuance of NBS Standard Reference Materials are a likely possibility, other mechanisms such as sample interchange are also being explored.

Finally there has recently been more interest in alternate nuclear fuel cycles such as the thorium fuel cycle. In this area NBS has no suitable standards and a thorium assay standard would seem to be a necessity. Consequently in this year's program, work has been started to prepare a thorium metal SRM similar to the metal now issued for uranium. This reference material will be in the form of a stick of high purity thorium certified for assay only. If interest and demand grows, other standards in this area are possible.

Much of the above work is in the initial stages of development and most of the research necessary to the measurement and certification of these standards is being supported by funding from the NRC and DOE through the NBS Office of Measurements for Nuclear Safeguards. This research is a necessary base to the measurement and certification process. Once the competencies and facilities are established, NBS should be in a position to provide other SRMs as needed for nuclear safeguard measurement.

RECEIVED MAY 11, 1978.

4

Decision Analysis for Nuclear Safeguards

JAMES P. SHIPLEY

Los Alamos Scientific Laboratory of the University of California, P.O. Box 1663,
Los Alamos, NM 87545

Materials accounting for safeguarding special nuclear materials (SNM) usually brings to mind instrumentation and measurement techniques for obtaining information on SNM locations and
amounts. The emphasis frequently is on data collection, a broad,
highly developed field that includes instrument design and the
specification and operation of complete measurement systems (see
Refs. 1,2,3,4,5 and the references therein). Just as important
is the analysis of materials accounting data to detect diversion
of SNM or process upset conditions. This paper is most concerned
with examining some efficient methods for analyzing and interpreting safeguards data.

Materials accounting for SNM currently relies heavily on
material-balance accounting following perodic shutdown, cleanout, and physical inventory. The classical material balance
associated with this system is drawn around the entire facility
or a major portion of the process, and is formed by adding all
measured receipts to the initial measured inventory and subtracting all measured removals and the final measured inventory.
During periods of routine production, control of materials is
vested largely in administrative and process controls, augmented
by secure storage for discrete items.

Although conventional material-balance accounting is essential to safeguards control of nuclear material, it has inherent
limitations in sensitivity and timeliness. The first limitation results from measurement uncertainties that desensitize the
system to losses of trigger quantities of SNM for large-
throughput plants. The timeliness of traditional materials

accounting is limited by the frequency at which the physical inventory is taken. There are practical limits on how often a facility can shut down its process and still be productive.

These conventional methods can be augmented by <u>unit process accounting</u> in which the facility is partitioned into discrete accounting envelopes called unit process accounting areas. A unit process can be one or more chemical or physical processes, and is chosen on the basis of process logic and whether a material balance can be drawn around it. By dividing a facility into unit processes and measuring all significant material transfers, quantities of material much smaller than the total plant inventory can be controlled on a timely basis. Also, any discrepancies are localized to that portion of the process contained in the unit process accounting area.

Material balances drawn around such unit processes during the course of plant operation are called <u>dynamic material balances</u> to distinguish them from balances drawn after a cleanout and physical inventory. Ideally, the dynamic material balances would all be zero unless nuclear material had been stolen (diverted). In practice they never are, for two reasons. First, the measuring instruments always introduce errors, for example, random fluctuations from electronic noise, or instrument miscalibrations. Second, constraints on cost or impact on materials processing operations may dictate that not all components of a material balance be measured equally often; therefore, even if the measurements were exact, the material-balance values would not be zero until closed by additional measurements.

Use of dynamic materials accounting implies that the operator of the safeguards system may be inundated with materials accounting data. Furthermore, although these data contain much potentially useful information concerning both safeguards and process control, the significance of any isolated (set of) measurements is seldom readily apparent and may change from day to day depending on plant operating conditions. Thus, the safeguards system operator is presented with an overwhelmingly complex body of information from which he must repeatedly determine the safeguards status of the plant. Clearly, it is imperative that he be assisted by a coherent, logical framework of tools that address these problems.

<u>Decision analysis</u> (6,7,8,9) is such a framework, and is well suited for statistical treatment of the imperfect dynamic material-balance data that become available sequentially in time. Its primary goals are (1) detection of the event(s) that SNM has been diverted, (2) estimation of the amount(s) diverted, and (3) determination of the significance of the estimates.

DECISION ANALYSIS

Decision analysis combines techniques from estimation theory and hypothesis testing, or decision theory, with systems analysis tools for treating complex, dynamic problems. The decision-analysis framework is general enough to allow a wide range in the level of sophistication in examining nuclear materials accounting data, while providing guidelines for the development and application of a variety of powerful methods.

The decision-analysis process is illustrated in Fig. 1. For nuclear materials accounting, the observed source generates true (error-free) data according to the switch position, which is determined by some unknown factor. The observed source is the nuclear materials processing line, and the unknown factor could be a divertor, for example. If the divertor is not stealing nuclear material, the switch is in the upper position; if diversion is occurring, the switch is in the lower position. A major part of the nuclear materials accounting problem is to choose between the two situations; the two choices are referred to as H_0, the null hypothesis under which no diversion has occurred, and H_1, the alternative hypothesis that diversion has occurred.

It should be noted that other factors besides diversion may cause SNM to be missing, which would appear as diversion. Part of the decision process consists of further investigations to discriminate among possible causes. For the purposes of this paper, no distinction is made between diversion and (apparently) missing material.

The true materials accounting data from the observed source, or equivalently the hypotheses H_0 and H_1, are not observed directly; otherwise, the decision problem would be trivial. Imperfect measurement devices (part of the data-collection function) provide corrupted data for the decision process. If measurement errors can be treated probabilistically, the resulting error statistics can aid subsequent analysis.

The estimation part of the analysis function is designed to take advantage of information in addition to that available in the measured data, with the objectives of obtaining more accurate and precise estimates of diverted material. If no other information on the observed source is available, the estimation algorithm simply passes the measured data and error statistics on to the decision function, along with the implicit assumption that when H_0 is true the material-balance values are all zero. Otherwise, estimation is based on more complicated models of source behavior, and estimate calculations assuming each hypothesis true are performed separately.

The source models for H_0 and H_1 true in Fig. 1 represent the translation into mathematical terms of whatever additional information exists concerning the source. Accurate and precise model construction is extremely important; inaccurate models cause incorrect, or biased, decisions, whereas imprecise models

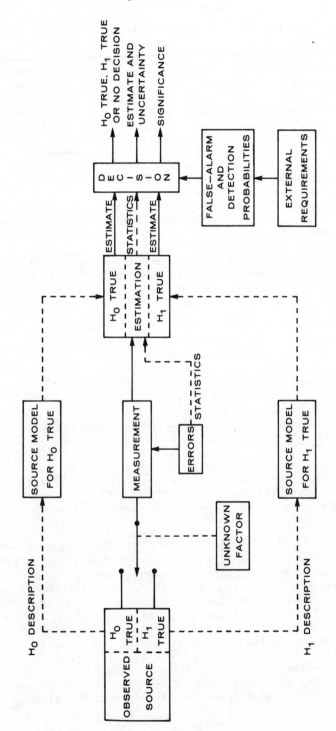

Figure 1. Structure of the decision analysis process

make it difficult to reach a decision having reasonable signifi-cance.

The decision function may be as informal as a perusal of the estimation results, or as structured as a statistical test with parameters fixed by administrative fiat. For the practical problems of nuclear materials accounting, a middle-ground approach is appropriate. A battery of statistical tests will facilitate quantified decision making, help eliminate personal biases, and form the basis for effective regulation. However, application of the tests and choice of test parameters should not be rigid or arbitrary; unforeseen circumstances and the pos-sibility of hidden errors require flexibility and subjective guidance in the decision process.

Although many different statistical tests are suitable for use in the decision process, they all have several characteris-tics in common. Each operates on the estimation results to decide whether H_0 or H_1 is true, and each requires some indication of desirable false-alarm and detection probabilities. One useful kind of test compares a likelihood ratio to a thresh-old, the likelihood ratio being defined roughly as the ratio of the probability that H_1 is true to the probability that H_0 is true, with the threshold determined by the desired false-alarm and detection probabilities. The variety of tests available to the decision process allows a wide range of tradeoffs among com-plexity, effectiveness, and applicability to special situations.

Decision analysis based on mathematically derived decision functions is appealing because it can quantify intuitive feelings and condense large collections of data to a smaller set of more easily understood descriptors, or statistics. It can also elimi-nate personal biases and other errors caused by subjective evalu-ation of data, while providing a degree of consistency for the decision process.

However, decision analysis should be considered as a manage-ment tool, not a management substitute. Unreasoning faith in test results is shortsighted for several reasons, the primary one being the inherent inadequacies of any tractable, mathemati-cal formulation of the hypotheses (i.e., the statistical models). In other words, statistical treatments are always based on sim-plified models derived from sometimes hidden assumptions that may not be valid for a particular situation, and possible effects on the decision process must be continually assessed.

A related problem is that a particular test can be defeated by choosing a diversion scheme that does not match the statis-tical model. A battery of tests and variable testing procedures reduce the probability of success of such schemes, especially if the tests and procedures are unknown to the divertor. This approach also tends to suppress the "beat-the-system" attitude exhibited sometimes, which is fostered by rigid application and interpretation of statistical tests; it also provides well-characterized information on which to base decisions.

As with all statistical procedures, a degree of reasonableness must be exercised; some results have statistical significance but no practical significance, and vice versa. Materials accounting data should certainly be examined carefully using statistical techniques, but the conclusions should be tempered by practical experience and personal judgment. Therefore, decision analysis need not be regarded as leading to an irreversible decision, but rather as an information-gathering procedure aimed at modifying attitudes towards the hypotheses on the basis of experimental evidence.

PROBLEM STATEMENT

Materials accounting data generally consist of a set of in-process inventory measurements at discrete times, each denoted by I(k), and a set of net material transfer measurements between those times, each represented by T(k) for those transfers occurring between times k and k+1. These measurements would satisfy the continuity equation for conservation of mass if the measurements were exact and all inventories and transfers were measured. However, SNM quantities can never be measured exactly, and diversion of SNM may have occurred or there may be other unobserved sidestreams, preventing measurement of all SNM. Therefore, the measurements satisfy a modified continuity equation:

$$I(k+1) = I(k) + T(k) - M(k+1), \quad k = 0,1,2,\ldots, \quad (1)$$

where M(k+1) is the material imbalance at time k+1 caused by measurement errors, unmeasured SNM, and diversion. The quantity M(k+1), called the k+1st material balance, can be determined from directly measurable quantities by inverting Eq. 1:

$$M(k+1) = - I(k+1) + I(k) + T(k) \quad . \quad (2)$$

Clearly, M(k+1) is a random variable, and the sequence $\{M(i), i = 1,2,\ldots\}$ is a stochastic process having probabilistic properties dependent on the inventory and transfer measurements. For example, if the measurement errors are unbiased, then each M(i) has a mean value equal to the amount of missing (or extra) material at each time i. Further, if the measurement errors are Gaussian, then each M(i) also has a Gaussian distribution with variance equal to the sum of the variances of the measurement errors.

Note that Eq. 2 shows that consecutive material balances are correlated, even if individual measurements are not. The ending inventory measurement for one material balance is the beginning inventory measurement for the next, resulting in a negative component of correlation between balances. Other correlations between individual measurements (e.g., measurement biases caused by instrument calibration error) yield additional correlation components, which are usually positive, between material balances. A method for treating correlations will be discussed below.

In an actual situation, we collect the set of inventory measurements $\{I(k), k = 0,1,...,N\}$ for some time period during which $N(>0)$ material balances have been drawn, the corresponding set of transfer measurements $\{T(k), k = 0,1,...,N-1\}$, and some statistical information on the measurement errors. Denote the aggregation of these data by $Z(N)$. The decision problem is to determine by analyzing $Z(N)$ whether diversion has occurred during the time interval, to estimate the amount of diversion, and to draw some conclusions about the significance of the estimate.

THE LIKELIHOOD RATIO

Hypothesis testing ($\underline{10},\underline{11},12$) provides a logical method for analyzing $Z(N)$ for possible diversion. To proceed in a general way, we form the two mutually exclusive, exhaustive hypotheses

H_0: diversion has not occurred,

H_1: diversion has occurred.

In developing specific decision algorithms, more mathematically quantified statements about the hypotheses will be necessary, and the particular form of each test will be strongly dependent on the corresponding hypothesis statements. However, these vague statements are sufficient for the general development.

For any particular $Z(N)$ that is observed, diversion may or may not have occurred, so that if H_0 is true, $Z(N)$ has the probability density function

$p[Z(N)|H_0]$,

and if H_1 is true, $Z(N)$ has the probability density function

$p[Z(N)|H_1]$.

These two conditional density functions are called the likelihood functions for the hypotheses H_0 and H_1, respectively. The values of the likelihood functions for a particular $Z(N)$ are relative measures of the likelihood that $Z(N)$ is governed by one or the other density function, or in other words, that H_0 or H_1 is true.

In making the decision whether H_0 or H_1 is true, we may commit either of two errors: we may decide that H_0 is true when it is not (a miss), or we may decide H_1 is true when it is not (a false alarm). Let the probability of a miss be P_M, and the probability of a false alarm be P_F. Decision algorithms may be derived from several different criteria concerning the selection of P_F and P_M. One of the most common criteria is to fix P_F and minimize P_M, which is known as the <u>Neyman-Pearson criterion</u>. Another method is to assign costs to incorrect decisions and minimize the expected value of the total cost of a decision. This criterion is known as the <u>Bayes risk criterion</u>, and it requires estimates of the prior probabilities that H_0 and H_1 are true. Whichever criterion is chosen, the decision test reduces to comparing the <u>likelihood ratio</u>, $L[Z(N)]$, to a threshold; i.e.,

$$\text{If } L[Z(N)] \triangleq \frac{p[Z(N)|H_1]}{p[Z(N)|H_0]} \quad \begin{cases} < T, \text{ accept } H_0 , \\ \geq T, \text{ accept } H_1 , \end{cases} \qquad (3)$$

where T is a threshold dependent on the criterion chosen. Roughly, Eq. 3 says that if $Z(N)$ is "enough" more likely to have occurred as a result of H_0 being true than of H_1 being true, then decide that H_0 is true; otherwise, decide that H_1 is true.

SEQUENTIAL DECISIONS

So far, the likelihood ratio test, Eq. 3, has been formulated as a fixed-length test; that is, all the data $Z(N)$ is collected before the test is performed. In actual practice, however, the optimum length and the proper starting point for the test will be unknown beforehand because the pattern of diversion, which is also unknown, is a determining factor in test characteristics. Furthermore, the materials accounting data naturally appear sequentially in time so that a sequential test procedure that selects its own length and starts from all possible initial points is appropriate. Such tests can be shown to require fewer data points on the average than fixed-length tests having the same characteristics (10,13).

Because the sequential likelihood ratio test, or sequential probability ratio test (SPRT), determines its own length, there are three possible results at each time, rather than two as in Eq. 3. The form of the SPRT is

$$\text{If } L[Z(N)] \quad \begin{cases} \leq T_0, \text{ accept } H_0 , \\ \geq T_1, \text{ accept } H_1 , \\ \text{otherwise, take another observation,} \end{cases} \qquad (4)$$

and the test is repeated for all possible starting points. As already discussed, the thresholds T_0 and T_1 can be found from a number of criteria, but some necessary information may be unavailable, making this approach less effective.

Given the false-alarm and miss probabilities, P_F and P_M, respectively, let the thresholds be defined by

$$T_0 = \frac{P_M}{1 - P_F} \, ,$$

$$T_1 = \frac{1 - P_M}{P_F} \, . \tag{5}$$

These are approximations, first devised by Wald (13), that can be shown to be conservative in the sense that use of T_0 and T_1 in a test will result in actual false-alarm and miss probabilities that are no larger than those originally given.

COMPOSITE HYPOTHESES

In many problems, one or both hypotheses may result in likelihood functions that contain an unknown parameter y; such a hypothesis is called composite. For example, y might be the (unknown) mean value of the observations. Without a value for y, the likelihood ratio cannot be calculated. One possible approach is to use estimates of y, under the corresponding hypotheses, for the actual y and proceed with the test. The most common estimate is the maximum likelihood estimate found by maximizing the appropriate likelihood function with respect to the unknown parameter. The resulting generalized likelihood ratio is

$$L[Z(N)] = \frac{\max\limits_{Y_1} p[Z(N)|H_1,y]}{\max\limits_{Y_0} p[Z(N)|H_0,y]} \, , \tag{6}$$

where Y_0 and Y_1 are the spaces of allowable values for y under the hypotheses H_0 and H_1, respectively (10,11,12).

SOME DECISION TESTS

As indicated above, the formulation of specific decision tests depends on more mathematically precise statements of the hypotheses. In particular, we seek statements of hypotheses that allow us to condense the quantity $Z(N)$ to one number $S(N)$ without loss of information. The number $S(N)$ is called a suffi-cient statistic (10) and is equivalent to knowledge of $Z(N)$. If such a $S(N)$ exists (which is usually true for SNM accounting) and if its form and density function are known, then the SPRT, Eq. 4, can be replaced by

$$\text{If } L[S(N)] \overset{\Delta}{=} \frac{p[S(N)|H_1]}{p[S(N)|H_0]} \begin{cases} \leq T_0, \text{ accept } H_0 , \\ \geq T_1, \text{ accept } H_1 , \\ \text{otherwise, take another observation.} \end{cases}$$

This approach is appealing because, now, the density functions are univariate and much more tractable mathematically. However, the form of the sufficient statistic may not be readily apparent without algebraic reduction of the original likelihood ratio. Further, a guess about the form of $S(N)$ may lead to a test hav-ing less desirable properties. The technique of reducing the original likelihood ratio is more general and always yields an appropriate sufficient statistic whenever one exists.

For any decision problem, a large number of different tests may be found, depending on the hypothesis statements. Following are some that have proven useful for SNM accounting.

The One-State Kalman Filter Statistic

Assume that all measurement errors are Gaussian and additive, and let the hypotheses be represented by

$$H_0: \quad M(k) = M_0 + v_M(k) \quad , \quad M_0 \leq 0$$

$$k = 1, 2, \ldots, \qquad (7)$$

$$H_1: \quad M(k) = M_1 + v_M(k) \quad , \quad M_1 \geq 0$$

where $v_M(k)$ is the measurement error for the kth material bal-ance. Then, the likelihood functions at any time k become (10,13)

$$p[Z(k)|H_0] = p[M(1),M(2),\ldots,M(k)|H_0]$$

$$= \prod_{i=1}^{k} [2\pi V_M(i)]^{-1/2} \exp - \frac{[M(i) - M_0]^2}{2V_M(i)} \quad ,$$

$$(8)$$

$$p[Z(k)|H_1] = \prod_{i=1}^{k} [2\pi V_M(i)]^{-1/2} \exp - \frac{[M(i) - M_1]^2}{2V_M(i)} \quad ,$$

where $V_M(i)$ is the error variance for the ith material balance, $M(i)$. Note that the two likelihood functions have unknown parameters M_0 and M_1. From the hypothesis statements, we must have $M_0 \leq 0$ and $M_1 \geq 0$. Thus, the maximum likelihood estimates for M_0 and M_1 (from the section on composite hypotheses) are, respectively,

$$\hat{M}_0(k) = \min \{0, \hat{M}(k)\} \quad ,$$

$$(9)$$

$$\hat{M}_1(k) = \max \{0, \hat{M}(k)\} \quad ,$$

where $\hat{M}(k)$ is the one-state Kalman filter estimate (14,15) for the material balance at time k. The estimate can be calculated recursively from the equations

$$\hat{M}(k) = \hat{M}(k-1) + B(k)[M(k) - \hat{M}(k-1)],$$

$$B(k) = \frac{V(k-1)}{V(k-1) + V_M(k)} \quad ,$$

$$(10)$$

where $B(k)$ is called the filter gain, and $V(k-1)$ is the variance of the error in the estimate $\hat{M}(k-1)$. $V(k)$ is also given recursively by

$$V(k) = [1-B(k)]V(k-1) \quad .$$

$$(11)$$

Initial conditions for the equations are $\hat{M}(0) = 0$, $V(0) = \infty$, which indicate our lack of prior knowledge about diversion. (Actually, $V(k)$ as calculated is conservative in that it is too large as a result of neglecting the common inventory measurement in consecutive material balances. This problem is resolved in the section on correlations.) Much more detail about the Kalman filter can be found in (16) and (17).

When these results are substituted into Eq. 8 and the generalized likelihood ratio is formed, the decision test can be shown to reduce to

$$
\text{If } \frac{\hat{M}(k)}{\sqrt{V(k)}}
\begin{cases}
\leq - \sqrt{2\,|\ln T_0|} & , \text{ accept } H_0 \; , \\[2mm]
\geq + \sqrt{2\,|\ln T_1|} & , \text{ accept } H_1 \; , \\[2mm]
\text{otherwise, take another observation,}
\end{cases}
\tag{12}
$$

and the test is performed sequentially for all k (until termination) and for all possible starting points. The decision, then, is based on comparison of each material-balance estimate with the standard deviation of its error, the same procedure as used for examining individual material balances one at a time for evidence of diversion.

It should be clear from the hypotheses and resulting equations that $\hat{M}(k)$ is an estimate of the <u>average</u> amount of missing (if $\hat{M}(k) > 0$) or extra (if $\hat{M}(k) < 0$) material <u>per material balance</u>. However, this does not mean that, for the test to work properly, the actual diversion must have occurred as a constant amount stolen during each balance period. Even if all the diversion took place within one balance period, the filter will still calculate the correct average per balance over any time interval containing the diversion.

Implicit in the hypotheses statements is the assumption that the sequence $\{M(i), i = 1,2,\ldots,k\}$ and its associated error variances are equivalent to $Z(k)$, that is, that knowledge of the separate inventory and transfer components of the material balances is unimportant. This would be true, for example, if the inventory measurement errors were small compared to those of the transfers. In that case, the Kalman filter estimate can be shown to be optimal in the sense that it is the linear, minimum-variance, unbiased estimate whenever the measurement error probability densities are symmetric about their means (<u>16</u>); i.e., the Gaussian error assumption is not necessary for calculating the estimate.

The CUSUM Statistic

If the material-balance error variances are all constant, $V_M(k) = V_M$ for $k = 1,2,\ldots$, then solution of Eqs. 10 and 11 results in

$$
\hat{M}(k) = \frac{1}{k} \sum_{i=1}^{k} M(i) \; , \qquad\qquad V(k) = \frac{V_M}{k} \; .
\tag{13}
$$

Multiplying both sides of the first equation by k yields a new statistic called the CUSUM (cumulative summation) ([18],[19],[20],[21]):

$$\text{CUSUM } (k) = \sum_{i=1}^{k} M(i) \quad , \tag{14}$$

which has variance

$$V_C(k) = V_I(k) + V_I(0) + \sum_{i=0}^{k-1} V_T(i) \quad , \tag{15}$$

where $V_I(\cdot)$ and $V_T(\cdot)$ are the inventory and transfer measurement error variances, respectively. The CUSUM statistic is interesting, even if the material-balance error variances are not constant, because it is an estimate of the total amount of missing material at any time during the period of interest. It is generally not optimal in any sense, but it has a useful physical interpretation and has become quite common.

A development analogous to that for the one-state Kalman filter yields the following SPRT:

$$\text{If } \frac{\text{CUSUM } (k)}{\sqrt{V_C(k)}} \begin{cases} \leq - \sqrt{2\left| \ln T_0 \right|} & \text{, accept } H_0 \ , \\ \geq + \sqrt{2\left| \ln T_1 \right|} & \text{, accept } H_1 \ , \\ \text{otherwise, take another observation,} \end{cases} \tag{16}$$

which is the same form as Eq. 12 in that an estimate of diversion is compared to its standard deviation.

The Two-State Kalman Filter Statistic

If the assumption that the inventory measurement errors are small compared to the transfer measurement errors is not valid, then an approach devised by Pike and his coworkers ([22],[23],[24],[25]) will yield a material balance estimate having smaller variance than the one-state filter. The technique is to estimate both the material balance and the inventory, which means that the filter now has two state variables rather than one. In recursive form, the filter equations are

$$\hat{I}(k) = \hat{I}(k|k-1) + B_1(k)[I(k) - \hat{I}(k|k-1)] \quad,$$

$$\hat{M}(k) = \hat{M}(k-1) + B_2(k)[I(k) - \hat{I}(k|k-1)] \quad, \tag{17}$$

$$\hat{I}(k|k-1) = \hat{I}(k-1) + T(k-1) - \hat{M}(k-1) \quad,$$

and $\hat{I}(k)$ and $\hat{M}(k)$ are the inventory and material-balance estimates, respectively, at time k based on all information through
time k. The filter gains $B_1(k)$ and $B_2(k)$ are given by

$$B_1(k) = \frac{V_{\hat{I}}(k)}{V_I(k)} \quad, \qquad\qquad B_2(k) = \frac{V_{\hat{I}\hat{M}}(k)}{V_I(k)} \tag{18}$$

where $V_{\hat{I}}(k)$ and $V_{\hat{I}\hat{M}}(k)$ are respectively the inventory estimate error variance and the covariance between the inventory and
material-balance estimate errors. They are given recursively by

$$V_{\hat{I}}(k) = \frac{V_{\hat{I}}(k|k-1)V_I(k)}{V_{\hat{I}}(k|k-1) + V_I(k)}$$

$$\tag{19}$$

$$V_{\hat{I}\hat{M}}(k) = \frac{V_{\hat{I}\hat{M}}(k|k-1)V_I(k)}{V_{\hat{I}}(k|k-1) + V_I(k)} \quad,$$

with

$$V_{\hat{I}}(k|k-1) = V_{\hat{I}}(k-1) - 2V_{\hat{I}\hat{M}}(k-1) + V_{\hat{M}}(k-1) + V_T(k-1), \tag{20}$$

$$V_{\hat{I}\hat{M}}(k|k-1) = V_{\hat{I}\hat{M}}(k-1) - V_{\hat{M}}(k-1) \quad.$$

The material-balance error variance at time k is

$$V_{\hat{M}}(k) = V_{\hat{M}}(k-1) - \frac{V_{\hat{I}\hat{M}}^2(k|k-1)}{V_{\hat{I}}(k|k-1) + V_I(k)} \tag{21}$$

The filter is initiated with $\hat{I}(0) = I(0)$, $V_{\hat{I}}(0) = V_I(0)$,
$\hat{M}(0) = 0$, $V_M(0) = \infty$, as before.
  By a development similar to that for the one-state filter,
the SPRT can be shown to reduce to

$$\text{If } \frac{\hat{M}(k)}{\sqrt{V_{\hat{M}}(k)}} \begin{cases} \leq - \sqrt{2 \left| \ln T_0 \right|} & \text{, accept } H_0 \text{ ,} \\ \geq + \sqrt{2 \left| \ln T_1 \right|} & \text{, accept } H_1 \text{ ,} \\ \text{otherwise, take another observation.} \end{cases} \qquad (22)$$

Generally, this test will be more sensitive than Eq. 12 because the estimate error variance is smaller.

This formulation has two other advantages. First, it provides a better estimate of the inventory. Second, the effects of correlated material balances caused by the common inventory measurement have disappeared as a result of the filter structure. However, we have bought these advantages at the expense of complexity and information requirements.

Nonparametric Tests

All tests such as those just discussed are called parametric because they depend upon knowledge of the statistics of the measurement errors. They also happen to work best when the measurement errors are Gaussian, a quite common but by no means all-inclusive situation. If the measurement error statistics are unknown or non-Gaussian, then nonparametric sufficient statistics (26-32) may give better test results. In addition, nonparametric tests can provide independent support for the results of parametric tests even though nonparametric tests are generally less powerful than parametric ones under conditions for which the latter are designed.

The two most common nonparametric tests are the sign test and the Wilcoxon rank sum test. The sufficient statistic for the sign test is the total number of positive material balances. That for the Wilcoxon test is the sum of the ranks of the material balances. The rank of a material balance is its relative position index under a reordering of the material balances according to magnitude. Other, possibly more effective nonparametric tests are being investigated. Further discussion of nonparametric tests is beyond the scope of this paper.

CORRELATIONS

Consider first the problem of correlated measurements, in particular, correlated transfer measurements. The following simplified treatment is due primarily to Friedland (33,34,35). Let the actual net transfer $T^a(k)$ be represented by

$$T^a(k) = T(k) - v(k) - w(k) \text{ ,} \qquad (23)$$

where T(k) is the transfer measurement, v(k) is the random measurement error (i.e., $E[v(k)v(k+j)] = 0$ for all $j \neq 0$, and $E[\cdot]$
is the expectation operator), and w(k) is the so-called "systematic error." Let us assume that w(k) is a bias that arises from
instrument miscalibration, say, and thus is constant between
calibrations. Further assume that the (constant) w(k) resulting
from any calibration is a Gaussian random variable with mean
zero and variance V_w. Then w(k) can be represented recursively by the difference equation

$$w(k) = a\,w(k-1) + (1-a)u(k), \qquad k = 1,2,\ldots, \tag{24}$$

$$\text{where} \quad a = \begin{cases} 0 & \text{if a calibration just occurred,} \\[2mm] 1 & \text{if a calibration has not just occurred,} \end{cases}$$

and u(k) is a Gaussian random variable with mean zero and variance V_w equal to the covariance between transfer measurements. Equation 24 can be appended to the state equations for
either the one- or two-state Kalman filter, which will then
yield an estimate of the bias w(k). Any systematic error can be
treated in this fashion merely by increasing the order of the
filter, but knowledge of the systematic error statistics is
required.

One of Friedland's major results (<u>33</u>) is that the optimum
material balance estimate can be expressed as

$$\hat{M}(k) = \tilde{M}(k) + D\hat{w}(k) \tag{25}$$

where $\tilde{M}(k)$ is the bias-free estimate, computed as if there were
no bias, and D is related to the ratio of the covariance of $\tilde{M}(k)$
and $\hat{w}(k)$ to the variance of $\hat{w}(k)$. Thus, calculation of $\hat{M}(k)$ can
be decoupled from the bias estimate until the final step.

This kind of systematic error is an example of a positive
correlation, and failure to account for it has two deleterious
effects. First, the material-balance estimate is biased, possibly giving a biased decision. Second, the variance of the
material-balance estimate error appears to be smaller than is
actually the case. This may result in a decision that seems to
be better than it is.

Now consider material balances that are correlated (nega-
tively) through the common inventory measurement, as for the
one-state Kalman filter. Write the kth material balance as

$$M(k) = -\,I^a(k) + I^a(k-1) - v_I(k) + v_I(k-1) + T(k), \tag{26}$$

where $I^a(k)$ is the kth actual inventory and $v_I(k)$ is the kth
inventory measurement error with variance $V_I(k)$. Define

$$v_1(k) = + v_I(k-1) , \qquad v_2(k) = -v_I(k) . \qquad (27)$$

In recursive form, these equations are

$$v_1(k) = - v_2(k-1) , \qquad v_2(k) = -v_I(k) , \qquad (28)$$

where v_1 and v_2 are now considered as state variables just as is the material balance in the one-state filter. In analogy to the treatment of positive correlations, Eq. 28 can be appended to the state equations for the one-state filter (for the two-state filter there is no need), which then gives estimates, $\hat{v}_1$ and $\hat{v}_2$, of the inventory measurement errors. That is, this method of treating the negative correlations also generates improved inventory estimates, which are given by

$$\hat{I}(k-1|k) = I(k-1) - \hat{v}_1(k|k) ,$$
$$\qquad\qquad\qquad\qquad\qquad\qquad (29)$$
$$\hat{I}(k|k) = I(k) + \hat{v}_2(k|k) .$$

Notice that $\hat{I}(k|k)$ is the filtered estimate of the inventory at time k and is based on the first k inventory measurements. The estimate $\hat{I}(k-1|k)$ also uses the first k inventory measurements, but it is the lag-one, smoothed estimate of the inventory at time k-1.

A negative correlation such as this, contrasted to the positive ones treated above, causes no bias in the material balance estimate. However, it does result in a material balance error variance that appears larger than actual if the correlation is ignored. The sensitivity of the corresponding decision test would, therefore, be degraded.

TEST APPLICATION

Procedure

As discussed above, we seldom will know beforehand when diversion has started or how long it will last. Therefore, the decision tests must examine all possible, contiguous subsequences of the available materials accounting data (1,2,3,18). That is, if at some time we have N material balances, then there are N starting points for N possible sequences, all ending at the Nth, or current, material balance, and the sequence lengths range from N to 1. Because of the sequential application of the tests, sequences with ending points less than N have already been tested; those with ending points greater than N will be done if the tests do not terminate before then.

Another procedure that helps in interpreting the results of tests is to do the testing at several significance levels, or false-alarm probabilities. This is so because, in practice, the test thresholds are never exactly met; thus, the true significance of the data is obscured. Several thresholds corresponding to different false-alarm probabilities give at least a rough idea of the actual probability of a false alarm.

Displaying the Results

Of course, one of the results of most interest is the sufficient statistic. Common practice is to plot the sufficient statistic, with symmetric error bars of length twice the square root of its variance, vs the material balance number. The initial material balance and the total number of material balances are chosen arbitrarily, perhaps to correspond to the shift or campaign structure of the process. For example, if balances are drawn hourly, and a day consists of three shifts, then the initial material balance might be chosen as the first of the day, and the total number of material balances might be 24, covering three shifts. Remember, however, that this choice is for display purposes only; the actual testing procedure selects all possible initial points and sequence lengths, and any sufficient statistic may be displayed as seems appropriate.

The other important results are the outcomes of the tests, performed at the several significance levels. A new tool, called the alarm-sequence chart ($\underline{1},2,\underline{3},\underline{18}$), has been developed to display these results in compact and readable form. To generate the alarm-sequence chart, each sequence causing an alarm is assigned (1) a descriptor that classifies the alarm according to its false-alarm probability, and (2) a pair of integers (r_1, r_2) that are, respectively, the indexes of the initial and final material balance numbers in the sequence. It is also possible to define (r_1, r_2) as the sequence length and the final material balance number of the sequence. The two definitions are equivalent. The alarm-sequence chart is a point plot of r_1 vs r_2 for each sequence that caused an alarm, with the significance range of each point indicated by the plotting symbol. One possible correspondence of plotting symbol to significance is given in Table I. The symbol T denotes sequences of such low significance that it would be fruitless to examine extensions of them; the letter T indicates their termination points. It is always true that $r_1 \leq r_2$ so that all symbols lie to the right of the line $r_1 = r_2$ through the origin. Examples of these charts are shown in the section on results.

TABLE I

ALARM CLASSIFICATION FOR THE ALARM-SEQUENCE CHART

Classification (Plotting Symbol)	False-Alarm Probability
A	10^{-2} to 5×10^{-3}
B	5×10^{-3} to 10^{-3}
C	10^{-3} to 5×10^{-4}
D	5×10^{-4} to 10^{-4}
E	10^{-4} to 10^{-5}
F	$< 10^{-5}$
T	~ 0.5

AN EXAMPLE

The Process

To demonstrate the application of decision analysis, we present results from a study (2) of materials accounting in a nuclear fuel reprocessing plant similar to the Allied-General Nuclear Services (AGNS) chemical separations facility at Barnwell, South Carolina (BNFP). The BNFP (36) is designed to use the Purex process to recover uranium and plutonium from power-reactor spent fuels containing either enriched uranium oxide or mixed uranium-plutonium oxide. Nominal capacity is 1500 MT/yr of spent fuel, which is approximately equivalent to 50 kg/day of plutonium.

For a plant such as BNFP, the most desirable areas for mate- rials accounting would be those containing the largest amounts of plutonium in a form most attractive to the divertor. The plutonium at the head end of the process is not attractive be- cause it contains lethal concentrations of fission products and is diluted approximately 100-fold with uranium. However, after the 1B column, the bulk of the fission products have been removed and the uranium/plutonium ratio has been reduced to 2/1. From this point the plutonium becomes increasingly attractive as it proceeds through the process to the plutonium-nitrate storage tanks. Hence, this area, the plutonium purification process (PPP), was selected for design of a dynamic materials accounting system. A block diagram of the PPP is shown in Fig. 2. Typical values for concentrations and flow rates are given in Table II, and Table III lists nominal in-process inventories.

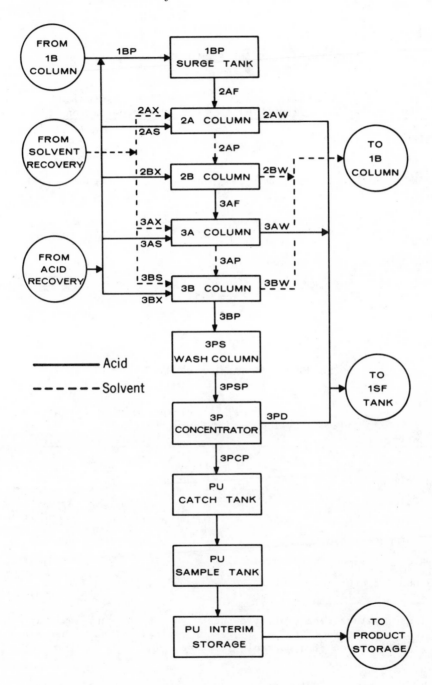

Figure 2. Block diagram of the plutonium purification process

TABLE II

CONCENTRATIONS AND FLOW RATES IN THE PPP

Stream	Flow (L/h)	Plutonium Concentration (g/L)
1BP	400	5
3PCP	8	250
2AW	500	trace
3AW	215	0.1
3PD	32	trace
2BW	150	trace
3BW	105	trace

TABLE III

IN-PROCESS HOLDUP IN TANKS AND VESSELS OF THE PPP[a]

Identification[b]	Volume (L)	Plutonium Concentration (g/L)	Plutonium Holdup (kg)
1BP Tank	1500	4.942	7.4
2A Column	700	c	4.6
2B Column	500	c	2.8
3A Column	600	c	5.4
3B Column	440	c	4.8
3PS Wash Column	20	58.70	1.2
3P Concentrator	60	250.	15.

---

[a] These values are not flowsheet values of any existing reprocessing facility but represent typical values within reasonable ranges of a workable flowsheet.

[b] See Fig. 2.

[c] A model of the concentration profiles and the holdup in the pulse columns is described in Ref. (2).

The Materials Accounting System

To isolate the PPP as a unit process requires flow and concentration measurements at the 1BP tank (input) and 3P concentrator (output). In addition, acid recyles (2AW, 3AW, 3PD) and organic recycle (2BW, 3BW) must be monitored for flow and concentration, and an estimate of the in-process inventory must be obtained. Table IV gives the required measurements and some possible measurement methods and associated uncertainties.

The relative precision of dynamic volume measurements is estimated to be 3% (1σ) for the 1BP tank, threefold more than for a conventional physical-inventory measurement because liquid is continuously flowing into and out of the tank during processing. Dynamic estimates of plutonium concentration in the 1BP and 3P concentrator tanks can be obtained from direct, in-line measurements (by absorption-edge densitometry, for example), or from combinations of adjacent accountability and process-control measurements.

TABLE IV

MATERIALS ACCOUNTING MEASUREMENTS FOR THE PPP

Measurement Point	Measurement Type	Instrument Precision (1σ, %)	Calibration Error (1σ, %)
1BP, 3PCP streams	Flow meter	1	0.5
	Absorption-edge densitometry	1	0.3
1BP surge tank	Volume	3	--
	Absorption-edge densitometry	3	--
2A,2B,3A,3B columns	See text	5-20	--
2AW,2BW,3AW, 3BW, 3PD streams	Flow meter	5	1
	NDA	10	2
3PS column	See text	5-20	--
3P concentrator	Volume (constant)	--	--
	Absorption-edge densitometry	1.5	--

High-quality measurements of the in-process plutonium inventory in the columns are the most difficult to make. In the reference design, the columns are fully instrumented for process control, including measurements of the position of the aqueous-organic interface and of the level and density of liquid in the product-disengagement volume. Much of the column holdup is in the product-disengagement volume, and a good measurement of this volume is available. However, only a crude estimate of plutonium concentration can be made without additional instrumentation. Relative precision for column-holdup measurements is estimated to be in the range of 5-20% (1σ). The 20% limit appears to be conservative in terms of discussions with industry and DOE personnel. A precision of 10% should be practicable using the current process-control instrumentation. Improvements toward the 5% figure (or better) will require additional research and development to identify optimum combinations of additional on-line instrumentation and improved models of column behavior.

Waste and recycle streams from the columns and the concentrator in the PPP are monitored by a combination of flowmeters and NDA-alpha detectors for plutonium concentration. The alpha monitors are already used for process control in the reference design and require only modest upgrading (primarily calibration and sensitivity studies) to be used for accountability as well. Flow measurements in the waste and recycle streams can be simple and relatively crude (5-10%) because the amount of plutonium is small. A rough calibration of the air lifts for liquid flow may suffice, or continuous level monitors in the appropriate head-pots could provide the required data.

Several measurement strategies have been investigated, including one in which material balances are drawn every hour, column inventory measurement precision is taken as 5%, and flow meters are recalibrated every 24 hours. This is the best of the strategies considered and is the one on which the following results are based.

Results

Decision analysis techniques have been used to evaluate the diversion sensitivity of this materials accounting strategy and others (2). Part of the evaluation consists of examining test results, in the form of estimate (sufficient statistic) and alarm-sequence charts, for various time intervals. Examples of typical one-day charts for the CUSUM and two-state Kalman filter, both with and without diversion, are given in Figs. 3 and 5; the horizontal marks indicate the values of the estimates, and the vertical lines are $\pm$ 1σ error bars about those estimates. The corresponding alarm-sequence charts are shown in Figs 4 and 6. The diversion level for the lower charts is 0.073 kg Pu/balance period, which is about 0.1 standard deviation of a single material balance. Results of a large number of tests show that the

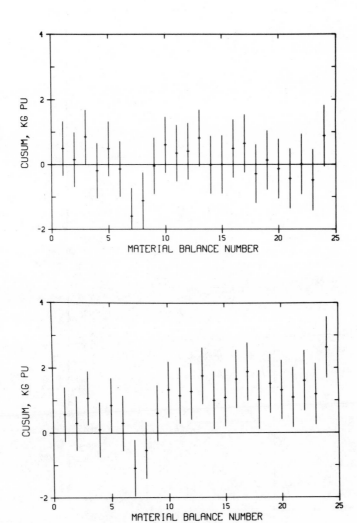

Figure 3. CUSUM charts without diversion (upper) and with diversion (lower)

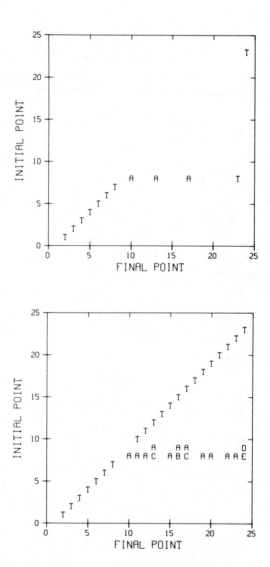

*Figure 4. Alarm-sequence charts for the CUSUMs
of Figure 3*

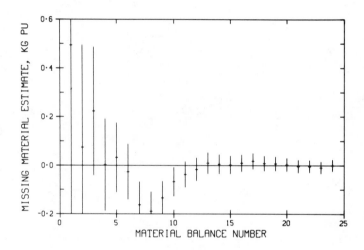

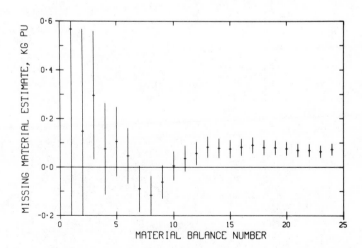

Figure 5. Kalman filter estimates of average missing material without diversion (upper) and with diversion (lower)

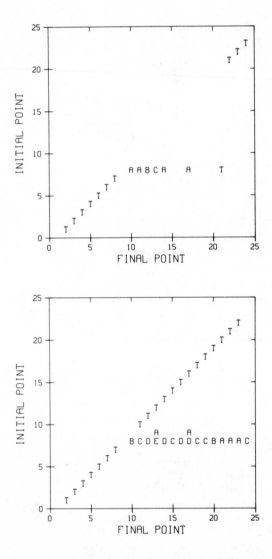

Figure 6. Alarm-sequence charts for the Kalman-filter estimates of Figure 5

two-state Kalman filter gives somewhat better results than the CUSUM, as expected.

In the course of evaluation, many such sets of charts are examined so that the random effects of measurement errors and normal process variability can be assessed; that is, we perform a Monte Carlo study to estimate the sensitivity to diversion. In applying decision analysis to data from a facility operating under actual conditions, only one set of data will be available for making decisions, rather than the multiple data streams generated from a simulation. In particular, direct comparison of charts with and without diversion, as shown here, will be impossible. The decision-maker will have to extrapolate from historical information and from careful process and measurement analysis to determine whether diversion has occurred.

The results of the evaluation are given in Table V. By comparison, current regulations require that the material-balance uncertainty be less than 1% (2σ) of throughput for each six-month accounting period, which corresponds to 75 kg of plutonium for this process. Such large improvement in diversion sensitivity is possible through the combination of timely measurements with the powerful statistical methods of decision analysis.

TABLE V

DIVERSION SENSITIVITY FOR THE PPP

Average Diversion per Balance (kg Pu)	Detection Time (h)	Total at Time of Detection (kg Pu)
2.6	1	2.6
0.075	24	1.8
0.025	168 (1 week)	4.2

LITERATURE CITED

1.   Shipley, J. P., Cobb, D. D., Dietz, R. J., Evans, M. L., Schelonka, E. P., Smith, D. B., and Walton, R. B., "Coordinated Safeguards for Materials Management in a Mixed-Oxide Fuel Facility," Los Alamos Scientific Laboratory report LA-6536 (February 1977).

2. Hakkila, E. A., Cobb, D. D., Dayem, H. A., Dietz, R. J.,
    Kern, E. A., Schelonka, E. P., Shipley, J. P., Smith, D.
    B., Augustson, R. H., and Barnes, J. W., "Coordinated
    Safeguards for Materials Management in a Fuel Reprocessing
    Plant," Los Alamos Scientific Laboratory report LA-6881
    (September 1977).

3. Dayem, H. A., Cobb, D. D., Dietz, R. J., Hakkila, E. A.,
    Kern, E. A., Shipley, J. P., Smith, D. B., and Bowersox, D.
    F., "Coordinated Safeguards for Materials Management in a
    Nitrate-to-Oxide Conversion Facility," Los Alamos
    Scientific Laboratory report LA-7011 (to be published).

4. Keepin, G. R., and Maraman, W. J., "Nondestructive Assay
    Technology and In-Plant Dynamic Materials Control--DYMAC,"
    in Safeguarding Nuclear Materials, Proc. Symp., Vienna,
    Oct. 20-24, 1975 (International Atomic Energy Agency,
    Vienna, 1976), Paper IAEA-SM-201/32, Vol. 1, pp. 304-320.

5. Augustson, R. H., "Development of In-Plant Real-Time
    Materials Control: The DYMAC Program," Proc. 17th Annual
    Meeting of the Institute of Nuclear Materials Management,
    Seattle, Washington, June 22-24, 1976.

6. Howard, R. A., "Decision Analysis: Pespectives on
    Inference, Decision, and Experimentation," Proc. IEEE,
    Special Issue on Detection Theory and Applications 58, No.
    5, 632-643 (1970).

7. Ref. (2), Vol. II, App. E.

8. Shipley, J. P., "Decision Analysis in Safeguarding Special
    Nuclear Material," Invited paper, Trans. Am. Nucl. Soc. 27,
    178 (1977).

9. Shipley, J. P., "Decision Analysis for Dynamic Accounting
    of Nuclear Material," paper presented at the American
    Nuclear Society Topical Meeting, Williamsburg, Virginia,
    May 15-17, 1978.

10. Sage, A. P. and Melsa, J. L., Estimation Theory with
    Applications to Communications and Control (McGraw-Hill,
    1971).

11. Lehmann, Testing Statistical Hypotheses (John Wiley and
    Sons, Inc., 1959).

12. Blackwell and Girshick, M. A., Theory of Games and
    Statistical Decisions (Wiley, 1954).

13. Wald, A., Sequential Analysis (John Wiley and Sons, Inc., 1947).

14. Kalman, R. E., "A New Approach to Linear Filtering and Prediction Problems," Trans. ASME J. Basic Eng. 82D, 34-45 (March 1960).

15. Kalman, R. E. and Bucy, R. S., "New Results in Linear Filtering and Prediction Theory," Trans. ASME J. Basic Eng. 83D, 95-108 (March 1961).

16. Meditch, J. S., Stochastic Optimal Linear Estimation and Control (McGraw-Hill, 1969).

17. Jazwinski, A. H., Stochastic Processes and Filtering Theory (Academic Press, 1970).

18. Cobb, D. D., Smith, D. B., and Shipley, J. P., "Cumulative Sum Charts in Safeguarding Special Nuclear Materials," submitted to Technometrics (December 1976).

19. Duncan, A. J., Quality Control and Industrial Statistics (R. D. Irwin, Inc., 1965).

20. Page, E. S., "Cumulative Sum Charts," Technometrics 3, No. 1, 1-9 (February 1961).

21. Evans, W. D., "When and How to Use Cu-Sum Charts," Technometrics 5, No. 1, 1-22 (February 1963).

22. Pike, D. H., Morrison, G. W., and Holland, C. W., "Linear Filtering Applied to Safeguards of Nuclear Material," Trans. Amer. Nucl. Soc. 22, 143-144 (1975).

23. Pike, D. H., Morrison, G. W., and Holland, C. W., "A Comparison of Several Kalman Filter Models for Establishing MUF," Trans. Amer. Nucl. Soc. 23, 267-268 (1976).

24. Pike, D. H. and Morrison, G. W., "A New Approach to Safeguards Accounting," Oak Ridge National Laboratory report ORNL/CSD/TM-25 (March 1977).

25. Pike, D. H. and Morrison, G. W., "A New Approach to Safeguards Accounting," Nucl. Mater. Manage. VI, No. 3, 641-658 (1977).

26. Thomas, J. B., "Nonparametric Detection," Proc. IEEE, Special Issue on Detection Theory and Applications 58, No. 5, 623-631 (May 1970).

27. Hajek, J. and Sidak, Z., Theory of Rank Tests (Academic
     Press, 1967).

28. Carlyle, J. W. and Thomas, J. B., "On Nonparametric Signal
     Detectors," IEEE Trans. Info. Theory IT-10, No. 2, 146-152
     (1964).

29. Tantaratana, S. and Thomas, J. B., "On Sequential Sign
     Detection of a Constant Signal," IEEE Trans. Info. Theory
     IT-23, No. 3, 304-315 (May 1977).

30. Capon, J., "A Nonparametric Technique for the Detection of
     a Constant Signal in Additive Noise," 1959 IRE WESCON
     Convention Record, Part 4, San Francisco, August 1959.

31. Gibson, J. D. and Melsa, J. L., Introduction to
     Nonparametric Detection with Applications (Academic Press,
     1975).

32. Puri, M. L. and Sen, P. K., Nonparametric Methods in
     Multivariate Analysis (Wiley, 1971).

33. Friedland, B., "Treatment of Bias in Recursive Filtering,"
     IEEE Trans. Autom. Contr. AC-14, No. 4, 359-367 (1969).

34. Friedland, B., "Recursive Filtering in the Presence of
     Biases with Irreducible Uncertainty," IEEE Trans. Autom.
     Contr. AC-21, No. 5, 789-790 (1976).

35. Friedland, B., "On the Calibration Problem," IEEE Trans.
     Autom. Contr. AC-22, No. 6, 899-905 (1977).

36. "Barnwell Nuclear Fuel Plant-Separation Facility Final
     Safety Analysis Report," Allied General Nuclear Services,
     Barnwell, South Carolina (October 1975).

RECEIVED JUNE 9, 1978.

A Nonlinear Method for Including the Mass Uncertainty of Standards and the System Measurement Errors in the Fitting of Calibration Curves[1]

W. L. PICKLES, J. W. McCLURE, and R. H. HOWELL

Lawrence Livermore Laboratory, Livermore, CA 94550

The goal of the work described here was to be able to produce highly accurate (0.1 to 0.2%) calibration curves of nondestructive assay instruments where the accuracy of the standards available is the limiting factor, or at least a major source of calibration error.

In reaching the ultimate accuracies possible for a particular NDA measurement system the instrument long-term precision is often not the limiting factor. The variability of sample preparation and the accuracy and applicability of the standards used for calibration of the instrument usually create the greatest source of uncertainty (1).

We have developed a mathematical method of dealing with these types of errors in a statistically correct way. Our first test of this method was with standards accuracy for x-ray fluorescense analysis of freeze-dried (2) UNO_3. The method can also be used to evaluate the importance of sample variability errors. The type of computer code we have used in this method is commercially available from several sources (3,4) as a package which requires only a small amount of input-output user generated software.

Method

Our LLL XRFA system (5) has a repeatable precision which has been measured to be 0.1% (two standard deviations). In attempting to utilize this system for accountability measurements in the nuclear fuel cycle, we were continually frustrated by the lack of high accuracy solid samples in the mass range from 10 to 1000 µg. We were finally able to produce UNO_3 standards by a freeze-drying method with an NBS traceable accuracy of 0.2% (one standard deviation) (1). These samples were thought to have particle size absorption, but because of the uniform fibrous nature (1) of the freeze-dried samples it was expected that these absorption effects would be calculable to high accuracy. We have used 100 of these standards to calibrate our XRFA instrument.

Since the mass accuracy error of the standards was estimated to be twice as large as the instrument precision errors, we felt

it was particularly important to include the mass uncertainty in
the calibration procedure. Our approach was to treat the mass
values of the standards in exactly the same way as we normally
treat the instrument's response to those standards. That is, the
mass value of each standard is a gravimetrically measured quantity.
The gravimetrical mass value, M_i, is not the "true mass" of the
standard. It differs from the true mass in a normal way. The
gravimetrically measured mass, M_i, has a 67% probability of
deviating from the true mass value by less than 0.2%. We therefore created a set of parameters which represent the true mass
values,

$$\left\{ X_i \right\}$$

There is one X_i, or true mass, for each standard. It is now
possible to use these new parameters in expressing the instrument
response calibration curve equation, YFUN.

$$YFUN = G(A,B,C,\mu_1,\mu_2,X_i)$$

The true mass X_i of the standard is one of the variables in the
calibration function instead of being a fixed constant. Consequently, the true mass, X_i, may be fit along with A,B,C,μ_1 and
μ_2, the "usual" calibration curve fit parameters.

    The result of this technique is to start from a set of gravi-
metrically measured standard mass values and measured XRFA instrument responses to those standards and arrive at <u>both</u> the most
probable, or true mass, of the standard, and the most probable
response value. This is diagrammed schematically in Figure 1.

    The fitting procedure is accomplished by a commercially
available (2), non-linear, unconstrained minimization, computer
program. The program minimizes the quantity chi-squared. Our
chi-squared not only involves the deviations in the instrument
response from the calibration curve as is usual, but must also
include the deviations of the gravimetric mass values from the
true mass. The value of chi-squared per degree of freedom is a
measure of the "goodness" of fit of the calibration curve and
true masses to all the experimental data. Our chi-squared is defined in Figure 2. The expression for chi-squared has two sums
of weighted, squared deviations. The first of these terms is
past. It is different in that the true mass, X_i, is used in place
of the gravimetrically measured mass, M_i. The second term is new,
and is the sum of the squares of the deviations of the measured
   s from the true mass, weighted by the gravimetric errors. The
 cual calibration curve function, YFUN, which we used in this
 ɔrk is shown in Figure 3. The function contains three terms;
 .he first term is a constant, the second is a term that represents
simple mass absorption, and the third term allows for absorption
in the long thin fibers of UNO_3 oriented perpendicular to the
plane of the sample. The fact that the free parameters in this

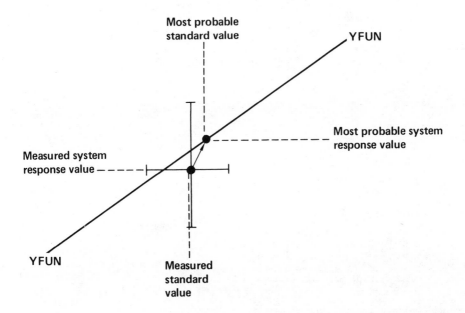

Figure 1. Overall result of nonlinear least squares fitting is a most probable system response value and a most probable standard mass value

System response

$$Q \equiv \sum_i \left[\frac{R_i - YFUN\,(A_1, B_2, \mu_1, \mu_2, C, \boxed{X_i})}{2\,\sigma_i} \right]^2 \quad \begin{array}{l}\text{Almost normal}\\ \text{CHISQUARED}\end{array}$$

Standards variations

$$+ \sum_i \left[\frac{\boxed{X_i} - M_i}{2\epsilon_i} \right]^2 \begin{array}{l}\text{New contribution}\\ \text{to CHISQUARED}\end{array}$$

Figure 2. New two-dimensional definition of chi-squared used in the nonlinear fitting technique. Note the use of true rather than gravimetric mass. To consistently use both types of errors chi-squared must include the standards mass errors.

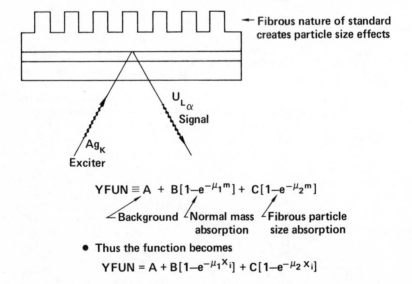

← Fibrous nature of standard creates particle size effects

U_{L_α} Signal

Ag_K
Exciter

$$YFUN \equiv A + B[1-e^{-\mu_1 m}] + C[1-e^{-\mu_2 m}]$$

∠Background ∠Normal mass   ∠Fibrous particle
                absorption   size absorption

• Thus the function becomes

$$YFUN = A + B[1-e^{-\mu_1 X_i}] + C[1-e^{-\mu_2 X_i}]$$

Figure 3. Actual calibration curve used in this work has a term for normal mass absorption and also a term for fibrous particle-size absorption. (For XRFA of freeze-dried UNO_3 standards we chose a physically realistic model.)

function $A, B, C, \mu_1, \mu_2, \{X_i\}$ appear as products and that the expression for chi-squared contains X_i's which are free parameters, dictates the use of a non-linear fitting program.

Results

The final results of using this technique is a "best-fit" value for A, B, C, μ_1, μ_2, and all the X_i's as shown in Figure 4 numerically and in Figure 5 graphically. As can be seen in Figure 5, 40% of the under response is due to simple mass absorption and 60% is due to particle size absorption.

Errors

A non-linear least squares fitting program does not use simple matrix inversion to obtain a unique best fit value for each free parameter and consequently does not produce a unique error matrix for the free parameters. However, estimates of the overall error is possible by two methods. In the first method, the curvature of chi-squared space near the best fit value of each parameter is an indication of the sensitivity of the fit to that parameter. The second and more useful method is to relax the errors on the gravimetric masses and/or the instrument response precision until a chi-squared per degree of freedom of approximately three is obtained. A chi-squared per degree of freedom of three means the probability that all the fit parameters are within one standard deviation of their "correct" value is 67%. We were able to obtain a chi-squared per degree of freedom of three by relaxing the instrument response errors to 0.2%. The conclusion we draw from this is we should accumulate counts on an unknown sample until the precision of the response is 0.1% and then the error we assign to the measurement of that sample will be 0.2% (1 sigma).

Summary

We have found that non-linear fitting techniques as described here to be a powerful method of creating realistic calibration curves for an NDA instrument and a particular standards set. The method uses both the gravimetric mass errors and the instrument response errors in a statistically consistent way. It incorporates the independent gravimetric measurement of the standards in the calibration curve parameters thus extracting all the experimental information available for the instrument response and the standards set. It determines the actual most probable value of each standard mass. It allows sensitive selection among the calibration curve models. It eliminates the need to cross measure standards, and it allows a realistic appraisal of the overall accuracy error of an NDA instrument and it's standards.

Abstract

At LLL we have used a sophisticated non-linear multiparameter fitting program to produce a best fit calibration curve for the response of an x-ray fluorescence analyzer to uranium nitrate,

- A set of most probable model parameters

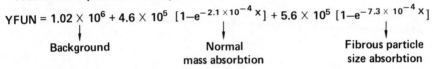

$$\text{YFUN} = 1.02 \times 10^6 + 4.6 \times 10^5 \; [1-e^{-2.1 \times 10^{-4} x}] + 5.6 \times 10^5 \; [1-e^{-7.3 \times 10^{-4} x}]$$

      Background   Normal   Fibrous particle
                           mass absorbtion   size absorbtion

- A set of most probable standard values

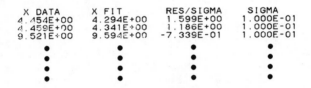

X DATA	X FIT	RES/SIGMA	SIGMA
4.454E+00	4.294E+00	1.599E+00	1.000E-01
4.459E+00	4.341E+00	1.186E+00	1.000E-01
9.521E+00	9.594E+00	-7.339E-01	1.000E-01

Figure 4. Actual numerical results of our fitting method showing our best fit parameters

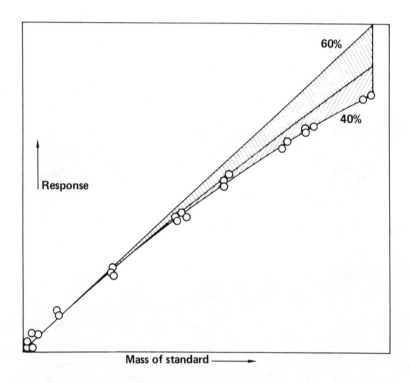

Figure 5. Graphical representation of our best fit calibration curve showing the normal mass absorption (40%) and the fibrous particle-size absorption (60%) under responses from the linear. Particle size and mass absorption are approximately equal.

freeze dried, 0.2% accurate, gravimetric standards. The program
is based on unconstrained minimization subroutine, VA02A. The
program considers the mass values of the gravimetric standards
as parameters to be fit along with the normal calibration curve
parameters. The fitting procedure weights with the system errors
and the mass errors in a consistent way. The resulting best fit
calibration curve parameters reflect the fact that the masses of
the standard samples are measured quantities with a known error.
Error estimates for the calibration curve parameters can be
obtained from the curvature of the "Chi-Squared Matrix" or from
error relaxation techniques. We have shown that non-dispersive
XRFA of 0.1 to 1 mg freeze-dried UNO_3 can have an accuracy of 0.2%
in 1000 sec.

[1]Work performed under the auspices of the U.S. Department of
Energy by the Lawrence Livermore Laboratory under contract number
W-7405-ENG-48.

Literature Cited

1. Pietri, C. E., Puller, J. S., Bingham, C. D., "The Chemical
    and Isotopic Analysis of Uranium, Plutonium, and Thorium in
    Nuclear Fuels," Proceedings of the ANS Topical Conference on
    Analytical Methods for Safeguard and Accountability Measure-
    ment of Special Nuclear Material, 1978.

2. Wong, C. M., Cate, J. L., Pickles, W. L., "Preparation of
    Uranium Standards for X-Ray Fluorescence Analysis," Proceed-
    ings of the ANS Topical Conference on Analytical Methods for
    Safeguard and Accountability Measurement of Special Nuclear
    Material, 1978.

3. International Mathematic and Science Library--XCSSQ Routine.

4. National Bureau of Economic Research--NL2SOL Routine.

5. Pickles, W. L., Cate, J. L., "Quantitative Nondispersive X-Ray
    Fluorescence Analysis of Highly Radioactive Samples for
    Uranium and Plutonium Concentration," Advances in X-Ray
    Analysis Vol. 17, Plenum Publishing Corp., New York, 1973.

RECEIVED JUNE 12, 1978.

Progress in the Verification of Reprocessing Input Analysis for Nuclear Material Safeguards

L. KOCH

Commission of the European Communities, Joint Research Centre, Karlsruhe Establishment, European Institute for Transuranium Elements, Karlsruhe, West Germany

E. MAINKA

Institut für Radiochemie, Kernforschungszentrum Karlsruhe, Postfach 22 66, D-7500 Karlsruhe 1, West Germany

The reprocessing input analysis is a key measurement in the nuclear fuel cycle not only from the safeguard point of view. The fuel can be directly analysed for the first time since its fabrication. A material balance between initial, fissioned and remaining material is possible with high accuracy. In addition, the portion of the fuel converted into special nuclear material e.g. plutonium is measured before it is purified for recycling.

Three possible methods for an input analysis are in use, of which the first one is applied usually by the plant operators, therefore the two latter ones can be used as a redundant measure (1,2,3).

The method of choice for most plants is the volume concentration method by which the volume of the dissolved fuel and the concentration of uranium and plutonium in it are measured.

The Pu/U-method needs only analysis of the ratio of these two elements in the spent fuel but requires information about the burn-up and the initial amount of the fuel.

A recently developed isotope correlation technique uses correlations between an isotopic ratio (which can be easily and with high accuracy measured) and the uranium or plutonium quantity. To obtain the input of the actinides the initial amount of the fuel has also to be known.

The following describes the progress made recently in Karlsruhe to verify the reprocessing input by automatic direct analysis, or by balancing pre- and post- irradiation amounts of fuel or by the isotope correlation technique.

AUTOMATES FOR DIRECT ANALYSIS

The automatic x-ray fluorescence spectrometer described earlier (4,5,6) has been improved: An instrument with a seven-channel analyser has been installed in the reprocessing facility, WAK, where it is manually operated under test. The automatic sample preparation stage has been completely redesigned. A more compact sampling device has been fabricated which is controlled by a microprocessor. This part of the automat is now being extensively tested cold under routine conditions.

A laboratory for automatic isotope dilution analysis, AIDA,
is in routine use for analysing samples taken by the EURATOM
safeguard inspectors. This laboratory comprises the following
automates:
- ion-exchange to condition uranium and plutonium for the subsequent isotope analysis (Fig. 1),
- a balance to weigh the sample and spike solution for the isotope dilution,
- an α-spectrometer to determine Pu-238 and transplutonite
  abundances,
- an automatic mass-spectrometer aided by a high-vacuum lock
  for continuous sample feeding (Fig. 2).

The conditioning of the reprocessing input solution sample
is done by sorption of the nitrato complexes of U and Pu on an
anion-exchanger and subsequent elution with dilute nitric acid
(7). The automat holds up to six columns each of which can be
programmed individually for the charging of two eluents and
for yielding up to four fractions with variable volumes (8).

The addition of known amounts of the spikes to a known aliquot of the sample solution is achieved by means of an automatic
balance which directly transmits the weights via a PDP 11/10 to
a IBM-370 computer for the analysis evaluation (Fig. 3) (9).

All operations of the mass-spectrometer necessary to ana-
lyse the isotopic abundance of uranium and plutonium are controlled by a dedicated computer. Continuous loading of the samples is achieved by a three-chamber high-vacuum lock which speeds
up the measurement by preheating the samples before they enter
the ion source of the instrument. From a thorough systematic
analysis of all operations used in manual operation mass spectrometry, elaborate software has been written for a dedicated
process computer. Starting from insertion of the samples into the
lock,the computer controls all steps such as baking out the
samples, heating up in the ion source, focussing and refocussing
the instrument, scanning at a pre-selected ion current with
possible subsequent measurement of uranium, plutonium and neo-
dymium from a single sample and reducing the mass spectra to
atom ratios, averaged over ten scans (9,10,11).

DATA PROCESSING

The handling of the data from routine reprocessing analysis,
generated for a single analysis over a period of several days,
justifies on-line data processing. The confidentiality and data
security required by safeguards make it mandatory. The system
employed in AIDA gives none of the operators direct access either
to the characteristics of the sample or to the result.

For each analysis the identity and characteristics of the
sample are filed in a specific matrix of the computer, which
periodically collects information available at the automates.
The data are sorted out according to the analysis indentity and
filed into the corresponding matrix. As soon as one matrix has

Figure 1. *Automatic device for six parallel U and Pu separations for subsequent MS analysis*

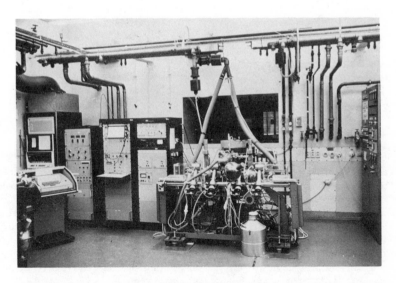

Figure 2. *Automatic MS and a high-vacuum lock for continuous sample feeding*

been completed the programme is selected according to the type
of data and the evaluation of the analysis starts (9).

The α-spectrometer and balance are on line to a PDP-11 which
reduces the source data and transmits them to the IBM-370. The
measurements of the mass spectrometer are converted to mass ratios by the process computer and are then transferred via the
PDP-11 to the IBM (Fig. 4) (12).

ISOTOPE CORRELATION TECHNIQUE

The results are filed in a DATABANK together with earlier
(historical) data where they can be checked by using the recently
developed isotope correlation technique. From a set of historical
data of similar origin, appropriate correlations are selected for
different applications.

The consistency of the new dataset is checked to see if the
analysis of one of the isotopes is faulty. By comparing different
correlations a possible error during the course of the analysis
can be spotted.

The burn-up determination needed for the Pu/U method can be
done either by the costly Nd-148 analysis (also by using AIDA) or
by employing correlations between e.g. plutonium isotope ratios
versus the burn-up (Tab. 1) (13,14,15,16).

A more advanced application of the isotope correlation technique is the prediction of the fissile material content directly
from isotope correlations. For this purpose, correlations are under study between the isotopic ratio within an actinide or
fission product and the fissile isotope content related to the
initial metal atoms of the unirradiated fuel (IMA). From such a
correlation the Pu-content could be deduced: Pu = Pu IMA $\cdot$ U_o
with U_o the initial amount of fuel and the Pu IMA correlation,
Pu IMA = a $\cdot$ Pu isotope ratio + b (Table 1).

CONCLUSIONS

The value of the automatic x-ray fluorescence spectrometery
lies in its potential for producing quickly a measurement of the
chemical concentration of fissile material, which is needed during inventory taking in the various hold-up vessels. At present,
tests are on the way to improve the accuracy by means of a multi-
channel analyser (18).

The developement of AIDA can be regarded as being completed.
Further automation would only be justified by larger sample
throughputs in order to reduce the analysis cost. From the present experience it can be concluded that AIDA has increased the
sample throughputs by five times compared to that from the earlier
manual operation. The accuracy (17) is not affected by the automation; on the contrary, the automatic procedure rejects samples
showing an odd behaviour in the mass spectrometer,thus eliminating
potential wrongly prepared material.A comparison between the three
above mentioned methods shows that the volume concentration procedure requires the highest number of measurements. However no additional information of the fuel history is required (Tab. 2). For
the second procedure the volume and density measurement as well as

Figure 3. Automatic balance for aliquotation used in isotope dilution analysis

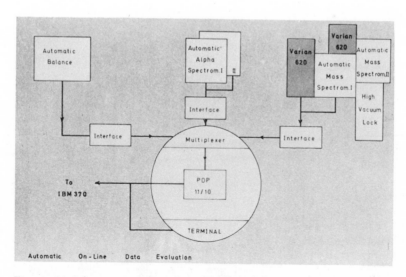

Figure 4. Scheme of on-line data handling between automates, process computers, and an IBM-370

Table I: Isotope correlations to predict burn-up (F_T), Pu-240 IMA or Pu-242 IMA. (Error: $s^2 = \sigma^2 (1 + 1/n)$; σ^2 = sums of squared deviations/(n-2); n = numbers of points; s = error of prediction; R = correlation coefficient)

$y = a \cdot$ (isotope ratio)	+ b	n	R	error %	reactor
Pu-240 = 2.00 E-3 (Xe-132/131)	− 2.97 E-3	19	.99	5.5	TRINO VERCELLESE
Pu-240 = 5.65 E-3 (Pu-240/239)	− 3.61 E-4	34	.98	3.2	OBRIGHEIM
Pu-240 = 7.80 E-3 (Pu-242/240)	+ 4.48 E-4	51	.98	3.9	GARIGLIANO
Pu-242 = 5.24 E-4 (Xe-132/131)	− 9.29 E-3	11	.99	6.6	OBRIGHEIM
Pu-242 = 1.89 E-3 (Pu-242/240)	− 5.98 E-4	51	.99	6.7	GARIGLIANO
F_T = 2.84 (Xe-132/131)	− 4.11	29	.98	5.8	all PWR
F_T = 14.3 (Pu-242/240)	− 0.436	78	.97	8.0	all PWR

Table II: Comparison of potential error sources (+) for each of the methods.

METHOD	MEASUREMENTS								OTHERS				ADDITIONAL INFORMATION	
ERROR SOURCES	Volume/Density	Shearing Losses	Undissolved	Dilution	Spiking	α & ms of Pu, U	ms of Nd (IDA)	ms of Xe, Kr	Aging effects	Cross Contamination	Unrepresentative sampling	Recycling of acid	Init. amount of fuel	Bank of historical data
Volume/concentration	+			+	+	+			+	+	+	+		
Pu/U ratio (^{148}Nd)	+	+			+	+	+		+	+	+	+	+	
(Corr)	+	+				+	+		+	+	+	+	+	+
Isotope Correlations (U,Pu)	+	+				+			+	+	+	+	+	+
(Xe,Kr)	+	+						+					+	+

the aliquoting of the sample is eliminated, but instead a burn-
up measurement is needed, which can be done by isotope correlations on actinide isotopic ratios thus avoiding the normally
applied, costly Nd-148 analysis. Information on the initial a-
mount of uranium is needed. The number of potential error sources is reduced compared to the first method. For the third procedure only a simple isotope analysis is required. However the
need for additional information on the fuel is increased. Besides the initial amount of uranium, historical data from which
isotope correlations can be deduced have to be provided. The potential error source of cross contamination during sampling inside the reprocessing plant is eliminated by using the fission
gas correlations, since a cross contamination from samples taken
from the exhaust gasses is unlikely.

1 Koch L., Bresesti M. Inst. of Nuclear Material Management, New
   Orleans, USA; Journal of the Institute of Nuclear Mat. Management, (1975) Vol. IV, No. III, P. 49
2 Koch L., Cottone G. Reaktortagung des dtsch. Atomforums, Karlsruhe, (1973) Tagungsberichte S. 287
3 Koch L., Ahrens H.J., Baeckmann A.v., Cricchio A., De Meester
   R., Romkowski M., van der Stijl E., Wilhelmi M. Intern. Atomic
   Energy Agency: Int. Symposium on Safeguarding of Nuclear Ma-
   terial, Wien; Proceedings of the Symposium (1975)
4 Baeckmann A.v., Koch L., Berg R. 70. Meeting of Americ. Chem.
   Soc. Chicago (1975)
5 Baeckmann A.v., Küchle M., Weitkamp C., Avenhaus R., Baumung
   K., Beyrich W., Böhnel K., Klunker J., Mainka E., Matussek P.,
   Michaelis W., Neuber J., Wertenbach H., Wilhelmi M., Woda H.,
   Hille F., Linder W., Schneider V.W., Stoll W., Koch L., Eberle
   R., Stegmann D., Zeller W., Krinninger H., Mausbeck H., Ruppert
   E. Intern. Atomica Agency, Genf (1971) Proceedings of 4. Confe-
   rence "Peaceful uses of Atomic Energy" A/Conf. 49/A/809
6 Baeckmann A.v., Koch L., Neuber J., Wilhelmi M. Intern. Atomic
   Energy Agency, Wien (1972) Proceedings SM-149/42
7 Koch L., Radiochim. Acta (1969) 12, 160
8 Bol D., Brandalise B., Bier A., De Rossi M., Koch L. EUR-5141
   (1974)
9 Brandalise B., Cottone G., Cricchio A., Gerin F., Koch L. EUR-
   5669 (1977)
10 Koch L., Wilhelmi M., Brandalise B., Rijkeboer C.,
   Romkowski M. Proc. of 7. Int. Mass Spectrometer Conf., Florence
   (1976) Vol. 7
11 Wilhelmi M., Brandalise B., Koch L., Rijkeboer C., Romkowski M.
   EUR 5504 d (1977)
12 Koch L. V Convegno di Spettrometria di Massa, Catania (1977)
   Annali de Chimica to be published
13 Ernstberger R., Koch L., Wellum R. ESARDA Symp. on Isotopic
   Correlation and its Application to the Nuclear Fuel Cycle,
   Stresa, Italy, May 9-11 (1978)

14 Wellum R., De Meester R., Kammerichs K., Koch L. ibid.
15 Brandalise B., Koch L., Rijkeboer C., Romkowski D. ibid.
16 Schoof S., Steinert H., Koch L. ibid.
17 Beyrich W., Drosselmeyer E. KFK-1905 (1975)
18 Neuber I., Flach S., Braun R., Stöckle D. in KFK-2465 P.3-8
   (1977)

RECEIVED JUNE 12, 1978.

7

Isotopic Safeguards Techniques[1]

C. L. TIMMERMAN

Battelle, Pacific Northwest Laboratories, Richland, WA 99352

The purpose of this paper is to explain and illustrate the idea and uses of isotopic safeguards techniques. The paper maintains a generalized, simple approach to facilitate understanding of the techniques. Once understood, the application, demonstration, and implementation of isotopic safeguards techniques becomes much easier.

Introduction

As uranium is burned in a nuclear reactor, plutonium is produced. Natural laws control this process, yielding simple relationships between the beginning and ending isotopic concentrations of the nuclear fuel. That is, the amount of plutonium produced is related to the amount of uranium remaining. The concentrations of many of the isotopes of both elements (U and Pu) can also be used to form useful relationships. These relationships, called isotopic functions, have become a prime tool in developing a reliable, straightforward method for verifying measurements of the total amount of plutonium and its isotopes produced in the nuclear fuel. The need to verify the amount of plutonium and other isotopics is illustrated by Figure 1. The fact that this technique would fill the material accounting safeguards gap in the fuel cycle between the fuel fabricator and the reprocessor is potentially its greatest asset. Isotopic safeguards has developed into a proven, reliable technique to verify the spent fuel content at the head end of a reprocessing facility.

The fact that the aforementioned relationships exist has long been recognized.(1,2) Both burnup experiments and calculational burnup codes have been used to study the transmutation of uranium to plutonium. The transmutation of isotopes is principally governed by simple first-order differential equations.(3) However, the coefficients of these equations depend on numerous core details and on the reactor operating history. The

[1]This work is currently sponsored by the Department of Energy through the International Safeguards Project Office, Brookhaven National Laboratory, under Control EY-76-C-06-1830. Much of the work was originally sponsored by The Arms Control and Disarmament Agency.

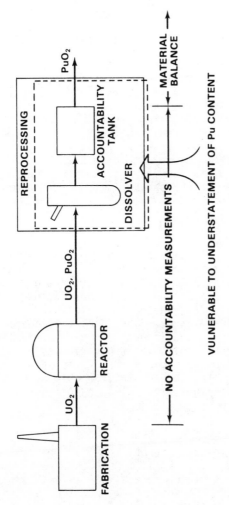

Figure 1. Why isotopic safeguards are important

complexity of the core models, the cross sections of the isotopes,
and the approximations needed to solve the equations has obscured
the simplicity of the relationships.

Similar relationships have been found in measurements of
isotopic concentrations in batches of dissolved spent fuel at
chemical reprocessing plants. These measurements represent
random samples of large amounts of irradiated fuel forming
batches as large as one tonne. Observations of these functional
relationships provide the best accuracy obtainable because they
represent actual spent fuel measurements coupled with predictions
from the theory associated with burnup calculations.

The isotopic functions have been investigated under programs
principally funded by the U.S. Arms Control and Disarmament
Agency in the past and currently by the Department of Energy
through the International Safeguards Project Office which is
managed by the Brookhaven National Laboratory. The purpose of
the programs has been to provide the International Atomic Energy
Agency (IAEA) with a practical verification technique for safeguarding the amount of plutonium input to chemical reprocessing
facilities. The programs are known as Isotopic Safeguards
Techniques.

Isotopic Functions and Ratios

An isotopic function is defined as a functional relationship
between the measured isotopic concentrations of uranium and
plutonium, or as the total elemental Pu/U mass ratio, for a given
dissolution batch. Examples of such functions include Pu/U
versus ^{235}U, ^{236}U versus ^{235}U, and $(^{239}Pu)^2$ versus ^{235}U. In
each example the Pu/U, ^{236}U, and $(^{239}Pu)^2$ are the dependent or
y variables and the ^{235}U is the independent or x variable.

The slope, or isotopic ratio, of two variables is defined as
the ratio of the dependent variable's irradiated value minus its
initial value divided by the independent variable's irradiated
value minus its initial value. The following example is provided
for illustration:

$$\text{Isotopic Ratio} = \frac{(^{236}U) \text{ irradiated} - (^{236}U) \text{ initial}}{(^{235}U) \text{ irradiated} - (^{235}U) \text{ initial}}$$

The above relationship is stated more simply as $\Delta\,^{236}U/\Delta\,^{235}U$ for
nomenclature convenience. Because the term $\Delta^{235}U$ is used so
frequently it has been further simplified to ^{235}D to indicate the
depletion of the ^{235}U isotope.

Obtaining a consistency for the slope of any function is of
primary importance to isotopic safeguards techniques. By having
a constant slope, the functions derived will be of a linear form.
The linearity provides a functional consistency of the isotopic
relationship over a large exposure or burnup range. This

functional consistency is necessary for an isotopic function to be of value to the technique and is a basic criterion for determining the merit of various isotopic functions. A graphical example of the linearity of the isotopic function ^{236}U versus ^{235}U is provided in Figure 2. This example illustrates the functional consistency for various enrichment groups of three different reactors.

An isotopic ratio can be used as a consistency check for batch-to-batch measurements in a reprocessing campaign. That is, the isotopic ratios formed from measurement data should be relatively constant for batches dissolved from the same reactor fuel lot. The ratio can also be used as a verification tool. Historical data from the same reactor, or a similar type reactor, can be used to form the isotopic ratio, which is compared (within accuracy limits) to incoming batch measurements from a present reprocessing campaign. In this way previous data can be used to verify present data through the consistency of isotopic ratios.

When historical data have been accumulated which highly characterize a particular fuel, these values may be applied to future similar fuels from the same reactor (assuming of course that controlling irradiation conditions are the same). The use of historical data to check successive discharges from the same reactor is illustrated in Table I for the Yankee Rowe reactor. All the Yankee fuel in this example is stainless steel clad and the same design. Reactor lots processed (10-20 tonnes U) at the chemical plant differ only in the enrichment and exposure. Two features of the empirical correlations are illustrated in Table I. First is the constancy of Pu/U versus ^{235}D for material of the same initial enrichment and design for a PWR. Second is the empirically derived expression relating the term $(Pu/U)/^{235}D$ to the initial enrichment. Each batch reported by the operator is tested by forming the ratio between Pu/U and ^{235}D. If all data check within statistical limits of the historical values and within direct IAEA measurements of input and product samples, the Pu/U ratio of input to the campaign is verified.

The case of no historical data but a high degree of consistency is illustrated by the data in Table II.($\underline{4}$) In this case, advantage is taken of the consistency of the data over a wide range of exposure, e.g., the Pu/U varies widely but the isotopic ratio is nearly constant. The general verification procedure outlined above is used. Each batch reported by the operator is tested to see if it differs from the majority of the data. If no significant differences are found, the verification of specified isotopic functions has been shown to a given accuracy.

Pu/U Ratio Method($\underline{5}$)

There are presently no adequate safeguards measures for checking the content of nuclear fuel from the fabrication facility through the reactor to input at a reprocessing plant.

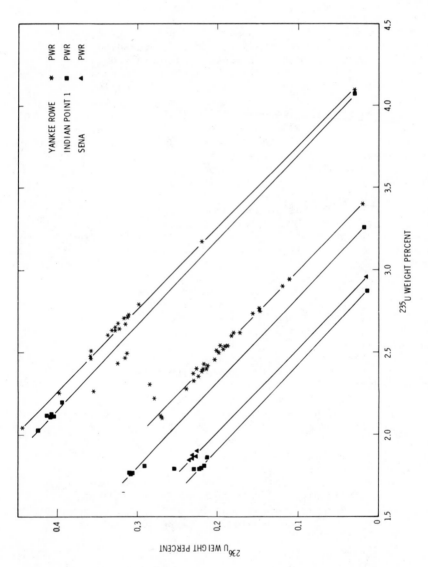

Figure 2. ^{236}U *weight percent as a function of* ^{235}U *weight percent*

Table I.
Comparison of Plutonium Results by Empirical
^{235}U Depletion Method to Measured Values
Yankee Rowe Fuel

No. of Dissolver Batches	^{235}Uo Wt. %	^{235}D	Grams Pu* Tonne U ^{235}D	Empirical Method Grams Pu Tonne U	Measured Grams Pu Tonne U	% Difference
16	3.404	.7981	5,727	4,571	4,530	0.91
14	3.406	1.0349	5,726	5,926	5,982	-0.94
11	4.101	1.5252	5,376	8,200	8,257	-0.69
7	4.101	1.4276	5,376	7,675	7,628	0.62
11	4.935	1.9820	4,957	9,824	9,899	-0.76
12	4.935	2.1594	4,957	10,703	10,692	0.10
11	4.941	2.0536	4,954	10,173	10,099	0.73

*Obtained by following equation:

$$\frac{\text{Grams Pu/Tonne U}}{^{235}D} = 7439 - 503 \ (^{235}Uo)$$

Table II (4).
Example Of Pu/U And Istopic Consistency
Yankee Rowe Core V Data

Raw Data

Input Batch #	Exposure MWd MTU	Total Pu g/TU	Uranium Isotopics		Plutonium Isotopics			
			^{235}U Wt. %	^{236}U Wt. %	^{239}U Wt. %	^{240}Pu Wt. %	^{241}Pu Wt. %	^{242}Pu Wt. %
1	9,700	4,968	3.176	0.2209	85.05	10.29	4.01	0.36
20	13,900	7,104	2.796	0.2997	79.35	12.53	6.65	0.78
10	15,800	7,930	2.675	0.3253	77.57	13.47	7.37	1.04
22	19,700	8,863	2.471	0.3589	75.14	14.54	8.29	1.34
18	21,400	9,794	2.254	0.3993	72.69	15.46	9.18	1.75
16	25,600	11,119	2.046	0.4454	69.31	16.34	10.85	2.50

Isotopic Ratios

Input Batch #	Exposure MWd MTU	Total Pu g/TU	Pu/U ^{235}D	% Dev.	Pu/U $\Delta^{236}U$	% Dev.	$\frac{^{240}Pu}{^{239}Pu \ (100 \ - \ ^{239}Pu)}$	% Dev.
1	9,700	4,968	5,371	-0.52	27,463	-0.01	7.748 E-3	2.76
20	13,900	7,104	5,448	0.91	27,334	-0.48	7.387 E-3	-2.04
10	15,800	7,930	5,417	0.33	27,516	0.19	7.486 E-3	-0.72
22	19,700	8,863	5,441	0.78	27,792	1.19	7.519 E-3	-0.28
18	21,400	9,794	5,306	-1.72	27,258	-0.75	7.535 E-3	-0.07
16	25,600	11,119	5,413	0.26	27,427	-0.14	7.567 E-3	0.35

$$\bar{X} = 5,399 \qquad \bar{X} = 27,465 \qquad \bar{X} = 7.540 \ E\text{-}3$$
$$\sigma = 0.98\% \qquad \sigma = 0.67\% \qquad \sigma = 1.58\%$$

Where: ^{235}D = wt. % ^{235}U final - wt. % ^{235}U initial

$\Delta^{236}U$ = wt. % ^{236}U final - wt. % ^{236}U initial

The need to verify the measured amount of plutonium in spent fuel at the input to a reprocessing plant is the basic reason for developing the isotopic safeguards concept. The plutonium-to-uranium ratio method is used to independently determine the plutonium content of a reprocessor's dissolver batch. The amount of plutonium at the input of the reprocessing plant is derived by the following equation:

Plutonium at Input = Pu/U ratio x (Initial Uranium - Burnup).

Measurements from three sources, two of them independent of chemical reprocessing, are used in the above equation. The initial uranium measured at fabrication must be measured accurately for the technique to be valid. Data(6) have proven this accuracy to be true, for the data have shown the fabricator's measurements to have a 0.15% relative precision and a 0.05% relative standard deviation of the systematic error. Correcting the initial uranium for burnup introduces the second independent source of information, the amount of heat produced by the uranium during irradiation. This calculated burnup is generally ±5% of the true value on an assembly level.(7) This error is not very large when one considers that only 2 to 5% of the initial uranium, depending on the burnup, is burned during irradiation in a reactor core. The third source of information is the Pu/U ratio obtained from the reprocessing plant. The accuracy of the Pu/U ratio as shown by our statistical studies is within 2% or less depending on the reactor and enrichment group. Using this ratio in conjunction with the final total uranium, obtained from the initial uranium and predicted burnup, a value for the total plutonium can be derived as stated in the above equation.

To obtain an independent value of the plutonium content using the Pu/U ratio method, it must be shown independently that the Pu/U ratios measured by the reprocessing plant are correct. By applying past similar reprocessing data and consistency checks to the Pu/U ratio and other isotopic ratios and functions, isotopic safeguards can provide an important complement and supplement to independent verification of input reprocessing batches.

It has been reported(8) that all of the observed differences are 1.0 to 1.5% or less for measured plutonium input compared to plutonium derived by isotopic safeguards methods. The 1.5% is considered a realistic upper limit, where larger differences on a batch level may require further sampling and investigation. As demonstration of the technique progresses, a definite practical upper limit can be determined.

A main process condition required to use the Pu/U ratio or other isotopic functions is that the input dissolver measurements must come strictly from the spent fuel from which they were derived. This is normally the case for head-end processes without recycle. If the acid which dissolves the spent fuel is recycled

from other sections of the reprocessing plant, corrections are
required for such additions of recycle acid. Recycle of any
nuclear material before measurement is made at the accountability
tank must be taken into account. This procedure of measuring all
recycle nuclear material holds for all verification procedures
and is not just particular to isotopic safeguards. In general,
verification procedures are easier to apply for the case of no
recycle because only one actual measurement needs to be made per
batch.

Recent Developments

A statistical package has been incorporated into the technique and is used to quantify the properties of various isotopic
functions and their associated data. The use of these automated
statistical procedures has provided many more meaningful results
which can be derived from the isotopic data.
The statistical tools used to analyze the data are namely
the paired comparison and regression analysis. A paired comparison results when a sample from a reprocessor's measurement batch
is analyzed by two different laboratories. These paired comparisons can be used to compare laboratories for bias, to estimate
measurement variances, and to indicate the presence of outlier
batch results. The second analysis tool, the linear regression,
initially involved the comparison of various linear regression
approaches. After much study of these various approaches, a
decision to use Deming's approach($\underline{9}$) was reached. Deming's
approach is a model in which errors in both variables (x and y)
are considered. Both an intercept model, $y = \alpha + \beta x$, and an
initial point model, $y - y_0 = \beta(x - x_0)$, are utilized to provide
the maximum statistical data from the regression analysis. For
example, the initial point model provides a stablizing line-
fitting method for data having a small burnup range, while the
intercept model supplies a more accurate fitting to larger ranged
data. The decision to use this regression technique with errors
in both the x and y variables resulted when it was shown($\underline{10}$)
that the random variances for the x and the y variable were
approximately equal, or at least that the error variance of y was
not substantially larger than the error variance of x. This fact
illustrated that errors occurred in both variables.
After choosing the appropriate statistical approach and
corresponding models, these statistics are useful in defining the
magnitude of the variance components along with identifying
anomalous or outlier data. The regression analysis also provides
a least squares type linear fitting of the data points grouped by
enrichment. This fitting supplies a more accurate estimate of
the slope and intercept than was previously possible. However,
possibly the most important feature of the statistics is that
they specify the accuracy basis for acceptable verification
results for each isotopic function.

Rough ideas had been used in the past to identify useful
linear functions. Because of the large number of isotopic
variables and functions possible, automated decision and identi-
fication techniques needed to be incorporated. The resultant
program determines the isotopic variables, and eliminates any
redundant functions that occur. The pairing of the variables
utilizes the idea that the relative change in one variable's
concentration over a sample burnup range should be approximately
equal (within 0.1%) to the relative change in the other variable's
concentration over the same burnup range in order to form a
linear function. Computer calculations have greatly improved the
efficiency and precision with which linear functions can now be
found. The following are examples of beneficial, linear functions
which have been studied by this method to date:

$^{240}Pu \times ^{241}Pu$	vs	$^{235}U \times ^{242}Pu$
^{235}U	vs	$^{239}Pu^2$
$^{239}Pu \times ^{240}Pu$	vs	$^{235}U \times ^{241}Pu$
$^{239}Pu^2 \times ^{240}Pu^2$	vs	^{240}Pu
$^{239}Pu \times (100 - ^{239}Pu)$	vs	$\Delta(^{235}U \times ^{239}Pu^2)$
^{240}Pu	vs	$\Delta(^{235}U^2)$
^{240}Pu	vs	$^{239}Pu \times (100 - ^{239}Pu)$
$^{239}Pu^2 \times (100 - ^{239}Pu)/^{235}U^2$	vs	$(100 - ^{239}Pu)$
Pu/U	vs	$(100 - ^{239}Pu)$
Pu/U	vs	^{235}U
^{236}U	vs	^{235}U

Data Base

A substantial data base has been collected to date. A
listing of the data sets according to type, reactor, and quantity
is provided in Table III. These data have increased the safeguards benefit of isotopic functions by providing a sufficiently
large set of data from which to work. Calculated burnup data are
included in the data base, which complement the empirical methods,
and this has provided greater theoretical insight into the technique. The features of the isotopic functions observed for the
measured data have also been observed for the calculated data,
thus supplying a high level of confidence in the isotopic
functions.

An in-depth analysis of the data base using the statistical
package and reactor operator and design information has led to a
labeling index as to the qualities and properties of every data
point in each data set. A list of the markings used for this

Table III.
Isotopic Safeguards Data Base

Reactors	Number of Input Batches Measured	Remeasured	Number Of Burnup Samples	Number of Sets Calculated Data
I. Pressurized Water				
1. Connecticut Yankee				1
2. Diablo Canyon				1
3. Fort Calhoun				1
4. Indian Point 1	20	20		1
5. Point Beach 1				1
6. San Onofre 1			6	3
7. Saxton			67	1
8. Sena	27	29		
9. Trino	30	48	23	3
10. Vver			5	
11. Yankee Rowe	89	82	35	5
12. H. B. Robinson 2			3	
Total	166	179	137	16
II. Graphite Moderated				
1. AGR				1
2. Calder Hall				1
3. Chapel Cross			24	
4. NPR	11			1
5. Windscale AGR	10			
Total	21		24	3
III. Boiling Water				
1. Big Rock Point 1	23			3
2. Browns Ferry 1				2
3. Dodewaard	13	6	13	
4. Dresden 1	62	18		3
5. Garigliano	31		18	
6. Humboldt Bay	20	20		3
7. KRB	15	29		1
8. LaCrosse				1
9. Nine Mile Point 1				1
10. Oyster Creek 1				1
11. VAK	3	4	10	1
12. JPDR-1			30	
Total	167	77	71	16
IV. Heavy Water				
1. NPD (Candu)	9	9	6	1
2. Douglas Point	3	6		1
3. Gentilly 1				1
4. Pickering				1
5. NRU			24	
Total	12	15	30	4
V. Fast Reactor				
1. EBR-11			18	
Total 35 Reactors	366	271	280	39 Sets

quality index is given in Table IV. It must be understood that most of the data falls under the ordinary category, for this is basically an index for marking and labeling unusual and outlier points to provide a clearer understanding into their meaning.

Our data base makes an effective safeguards tool in itself because of the isotopic functions now developed. Measurement histories of particular reactors and/or fuels become inherent as the data bank increases. Perhaps the most important aspect of the data base is that the isotopic functions determined from new reprocessing measurements should be consistent with past results and with properties defined by the data base for that type of reactor fuel. In this way the data base contributes in a positive way to the future analyses of reprocessing input measurement.

Table IV.

Quality Index Listing

Property

Normal data point

Exposure averaging

Enrichment averaging

Exposure and enrichment
averaging

Fringe effects

Exposure averaging and
fringe effects

Cladding differences

Unirradiated assembly

Uranium outlier

Plutonium outlier

94 NUCLEAR SAFEGUARDS ANALYSIS

NUCLEAR SAFEGUARDS ANALYSIS

I am indebted to Dean Christensen for his training and knowledge which has aided in my understanding of this technique. Kirk Stewart is also to be acknowledged for developing the statistics applicable to isotopic safeguards techniques.

94 NUCLEAR SAFEGUARDS ANALYSIS

94   NUCLEAR SAFEGUARDS ANALYSIS

Acknowledgment

I am indebted to Dean Christensen for his training and knowledge which has aided in my understanding of this technique. Kirk Stewart is also to be acknowledged for developing the statistics applicable to isotopic safeguards techniques.

Literature Cited

1. Rider, B. F., Russel, Jr., J. L., Harris, D. W., Peterson, Jr., J. P., "The Determination of Uranium Burnup in MWd/Ton," GEAP-3373, Vallecitos Atomic Laboratory, General Electric Company, Pleasanton, CA, March 1960.
2. Moeken, H. H. Ph. and Bokelund, H., "Verification of the Uranium and Plutonium Measurements on a Batch of Dissolved Reactor Fuel," ETR-235, European Company for Chemical Processing of Irradiated Fuels, Mol Belgium, June 1969. Available from NTIS.
3. Beets, C., "Role of Measurements of Nuclear Materials in Safeguards," presented at Symposium on Practical Applications of Research and Development in the Field of Safeguards, Rome, March 7-8, 1974. Centre d'Etude de L'Energie Nucleaire, Belgium.
4. Schneider, R. A., Stewart, K. B., and Christensen, D. E., "The Use of Isotopic Correlations in Verification (Safeguards Application)," BNWL-SA-4251, presented at IAEA Working Groups on the Use of Isotopic Composition Data in Safeguards, Vienna, Austria, April 10-14, 1972, Battelle, Pacific Northwest Laboratories, Richland, WA 99352, February 1972.
5. Stewart, K. B. and Schneider, R. A., "Properties of the Pu Estimate Based on Weighted Pu/U Values," IAEA SM-133/55, Safeguards Techniques, IAEA, Vienna, Austria, July 1970.
6. Stephens, F. B., et al., "Methods for the Accountability of Uranium Dioxide," NUREG-75/010, Lawrence Livermore Laboratory, June 1975.
7. Reactor Burn-up Physics, Proceedings of a Panel on Reactor Burn-Up Physics Organized by the International Atomic Energy Agency, Vienna, Austria, July 12-16, 1971, February 1973.
8. Christensen, D. E. and Schneider, R. A., "Summary of Experience with Heavy-Element Isotopic Correlations," Safeguarding Nuclear Materials, Vol. II (p. 377), IAEA, Vienna, Austria, 1976.
9. Deming, W. E., Statistical Adjustment of Data, Willey, New York, 1943.
10. Timmerman, C. L., Christensen, D. E., and Stewart, K. B., "Statistical Evaluation of Isotopic Safeguards Data," BNWL-SA-6371, presented at the 18th Annual Meeting of the Institute of Nuclear Materials Management, Washington, DC, July 1, 1976.

RECEIVED JUNE 6, 1978.

Nuclear Safeguards Applications of Energy-Dispersive Absorption Edge Densitometry

T. R. CANADA, D. G. LANGNER, and J. W. TAPE

Los Alamos Scientific Laboratory of the University of California, P.O. Box 1663, Los Alamos, NM 87545

For a number of years, absorption edge densitometry (x-ray absorption edge spectrometry) has been a somewhat under utilized tool of the analytical chemist (1-8). This reasonably matrix-independent technique, which requires the measurement of photon transmissions through a sample at two energies, one on each side of the elemental absorption edge of interest, has found in the past only limited application in the measurement of special nuclear materials (9,10,11). This is due primarily to the limited flexibility of the wave length dispersive spectrometry approach and its associated complex measurement procedures. High-resolution energy-dispersive spectrometers have alleviated, to a large degree, these drawbacks while maintaining the attractive matrix insensitivity feature of the technique.

This paper reviews briefly the theory of energy-dispersive absorption edge densitometry, describes its recent applications to the measurement of special nuclear materials (SNM) and discusses some possible future adaptations that will broaden its impact in the field of nuclear safeguards. Although the examples reviewed are concerned only with plutonium, uranium, and thorium, the extension of the technique to other elemental determinations is straightforward.

I. TECHNIQUE DESCRIPTION

The basic experimental components of energy-dispersive photon transmission measurements are shown schematically in Fig. 1. The ratio of photon intensities from the transmission source measured at an energy E with and without the sample present determines the transmission T, where

$$T = \exp(-\mu_s \rho_s x) \exp(-\mu_m \rho_m x).$$

x is the sample thickness, μ_s and μ_m are the mass attenuation coefficients at E, and ρ_s and ρ_m are the densities of the SNM of interest and the matrix materials (everything else), respectively. The collimation defines the detector-source geometry

and reduces the detector sensitivity to photons originating
within the sample, whether due to gamma-ray decay or
fluorescence.

An absorption edge densitometry-based assay requires that
photon transmission measurements be made through a sample at two
photon energies (12). The ratio of transmissions for two photon
beams at energies E_U and E_L through an SNM-bearing sample is
given by

$$T_U/T_L = \exp(-\Delta\mu_s \rho_s x) \, \exp(-\Delta\mu_m \rho_m x) \tag{1}$$

where $\Delta\mu_i = \mu_i^U - \mu_i^L$.

Figure 2 is a plot of μ versus photon energy for three
materials, water, tin, and uranium. The uranium curve shows a
sharp discontinuity at 116 keV corresponding to the uranium
K-absorption edge. It is this element-specific discontinuity
that is the basic signature on which absorption edge densitometry is based. The energy E at which a given elemental
absorption edge appears is in general unique, and in the
immediate neighborhood of E the mass absorption coefficients for
all other elements are monotonically decreasing functions of
energy.

If E_U and E_L of eq. (1) are picked to be E plus and
minus a ΔE (where ΔE is small), respectively, then $\Delta\mu_m \simeq 0$,
and ρ_s is given by

$$\rho_s \simeq -\ln R/\Delta\mu_s x \tag{2}$$

where $R = T_U/T_L$. As $\Delta E \to 0$, approximation (2) approaches an
equality, i.e., the measurement is specific to the SNM of
interest and is independent of the matrix properties. The
sensitivity of this technique is determined by the product $\Delta\mu_s x$
and may be increased by having a larger sample thickness x or
picking an absorption edge for which $\Delta\mu_s$ is larger. The uranium
and plutonium K and L_{III} absorption edges are at energies that
make transmission measurements practical. A list of these and
the corresponding $\Delta\mu$ values are given in Table I.

The relative precision with which R must be measured to give
a designated relative precision in ρ_s is given by

$$\frac{dR}{R} = (\Delta\mu_s \rho_s x)\frac{d\rho}{\rho} \tag{3}$$

The mean areal SNM density, $\rho_s x$ in g/cm^2, for which $\Delta\mu_s \rho_s x = 1$,
is a useful parameter to indicate the lower bounds of applicabil-
ity of a given absorption edge. At the uranium and plutonium

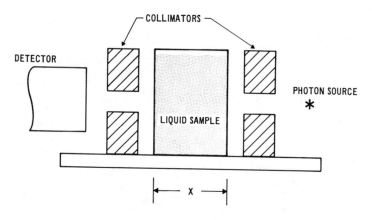

Figure 1. Basic experimental components of photon transmission measurements

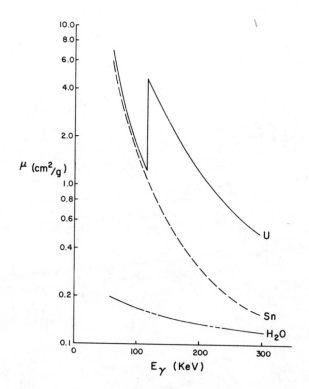

Figure 2. Total mass absorption coefficients as a function of energy for H_2O, Sn, and U

Table I. L_{III} and K absorption edge energies for uranium and plutonium.

Absorption Edge	Element	Edge Energy (keV)	$\Delta\mu$ (cm^2/gm)
L_{III}	U	17.17	54.60
	Pu	18.05	51.90
K	U	115.60	3.65
	Pu	121.76	3.39

K and L$_{III}$ edges these are approximately 0.3 g/cm^2 and 0.02
g/cm^2, respectively. Note that the sample thickness, x, is
limited by practical limitations upon the transmission source
intensity.

In general, high-energy resolution photon detectors, Si(Li),
GeLi(Li), and Ge, are required to achieve the accuracy, matrix
insensitivity, and versatility available in the application of
this technique. The choice of detector depends upon the
transmission source used for a given problem.

II. APPLICATIONS

A variety of transmission photon sources are available that
satisfy the criterion that ΔE approach zero. These include
x-ray generators, bremsstrahlung sources, and certain gamma-ray
emitting radioisotopes. The choice of source is dependent upon
a number of factors including cost, SNM density, design limitations, desired accuracy, and allowed assay time. The x-ray
generator is probably the most versatile source but also the
most expensive. Its use has many advantages: (a) the bremsstrahlung energy and intensity are variable, (b) the energy
displacement, ΔE, from the absorption edge is limited only by
the detector resolution, and (c) multiple simultaneous elemental
determinations, such as plutonium/uranium are possible.
Bremsstrahlung sources share advantages (b) and (c) and are in
addition compact and inexpensive. Radioisotopic gamma-ray
sources are only applicable when their decay schemes include
gamma rays near enough in energy to the absorption edge of
interest. However, they are convenient and inexpensive when
available with the correct energies.

A. Radioisotopic Sources

SNM-bearing samples with $\rho_s x \gtrsim 0.2$ g/cm^2 may be conveniently
assayed at the K-absorption edge using gamma-ray emitting
radioisotopes. Table II summarizes a number of source-gamma-ray
combinations that bracket the uranium and plutonium K-edges.

In the case of Pu, the ^{75}Se and ^{57}Co gamma rays at 121.12
keV and 122.06 keV, respectively, bracket the Pu K-edge at
121.795 keV so closely that, in most cases, approximation (2)
may be considered an equality (13). Figure 3 shows an example
of a measured calibration curve using these sources and a set of
2-cm-thick plutonium-bearing solutions. It also shows a similar
curve for the two ^{75}Se lines at 121.12 keV and 136.00 keV,
which, as a result of a larger ΔE from the edge, exhibits a
smaller slope and a larger matrix sensitivity, i.e., a measured
intercept> 1.

To demonstrate the matrix sensitivity for these ratios of
transmissions, measurements were made through a plutonium-
bearing solution (ρ_{Pu} = 67.5 g/ℓ) and a variety of tin thicknesses. The results are plotted in Fig. 4 as R versus the

Table II.   Convenient gamma rays for use in K-edge densi-
             tometry measurements of plutonium and uranium.

Plutonium
E_K = 121.8

Source	Ey(keV)	$\Delta\mu_{Pu}(cm^2/gm)$
^{75}Se	121.1	
	136.0	2.12
^{75}Se	121.1	
^{57}Co	122.1	3.38

Uranium
E_K = 115.6

Source	Ey(keV)	$\Delta\mu_{Pu}(cm^2/gm)$
^{169}Yb	109.8	
	130.5	2.20
^{169}Yb	109.8	
^{57}Co	122.1	2.70
^{182}Ta	113.7	
	116.4	3.53

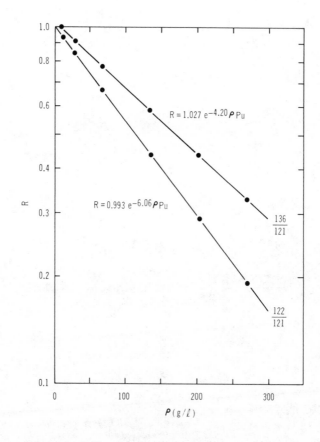

Figure 3. Calibration curves for the measurement of 2-cm thick Pu bearing solutions using the 121 keV and 122 keV gamma rays of ⁷⁵Se and ⁵⁷Co and using the 121 keV and 136 keV ⁷⁵Se gamma rays

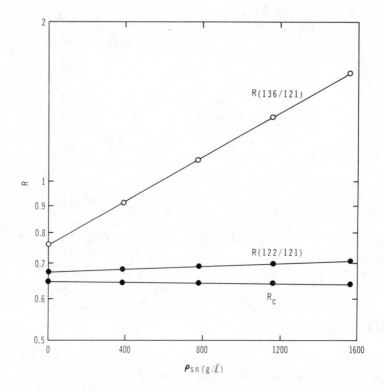

Figure 4. Ratio of transmissions, R, as a function of Sn matrix density for a 67.5 g/L Pu solution and two transmission energy combinations (R_c is the (136/121) ratio corrected for matrix)

effective tin density had it been distributed in the solution.
These data show the matrix sensitivity of R(122/121) to be
reasonably small whereas that at R(136/121) is significant.

It is possible to correct the measured R value when matrix
contaminants with Z > 50 are known or suspected to be present in
a sample. As shown in Fig. 5 (14), log (μ) versus the log (E)
may be approximated as a straight line over a limited energy
range. Figure 6 shows that the slope of this line, m, is
approximately a constant for all elements with Z > 50. Transmissions measured near the K-edge may then be extrapolated to
that which would have been measured at the absorption edge
giving (15)

$$\rho_s = \frac{-1}{\Delta\mu_s x}\left[\left(\frac{E}{E_U}\right)^{-m} \ln T_U - \left(\frac{E}{E_L}\right)^{-m} \ln T_L\right] \qquad (4)$$

where $\Delta\mu_s$ is measured at the absorption edge energy, E, and m ≈
2.55. Also shown in Fig. 4 is the ratio obtained from the
corrected transmission values, R (136/121), which varies by
only 1.5% with the addition of tin equal to ∿ 23 times the
plutonium concentration.

The 121 keV/122keV ratio is presently being incorporated
into a plutonium isotopic concentration spectrometer (16). The
feasibility experiments were conducted with 3.5-cm-thick
solutions containing 130-360 g Pu/ℓ. The transmission measurements were made simultaneously for both lines with a mixed ^{57}Co-
^{75}Se source and the resulting doublet computer fitted. The results showed an accuracy of < 0.5% but indicated a nonlinearity
in the calibration curve at the extremes of the concentration
range. The latter may be due to the imbalance in the relative
peak intensities at these concentrations and the corresponding
difficulties in fitting the doublet accurately.

No obvious gamma-ray lines exist that bracket the K-edge of
uranium as closely as do the 121.1- and 122.1-keV gamma rays for
that of plutonium. Figure 7 summarizes a series of measurements
made with 1-cm-thick solutions containing 50-400 g U/ℓ. These
data have been normalized to a 1-cm-thick water sample (the
uranium sample solvent) and thus do not reflect any matrix
sensitivity. The peak-to-background ratios for the Ta 116-keV
and 113-keV and the ^{169}Yb 118-keV gamma rays are too poor to be
useful in a practical assay instrument. An optimum combination
now appears to be a mixed ^{57}Co (122 keV) and ^{169}Yb (110 keV)
source.

The ^{169}Yb 109.8-130.9 keV gamma rays are presently used in a
Ge(Li) detector based solution measurement system (Fig. 8) to

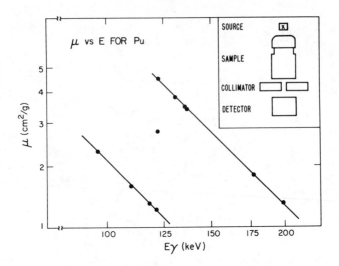

Figure 5.   Measured mass absorption coefficient of Pu vs.
energy plotted on a log–log scale

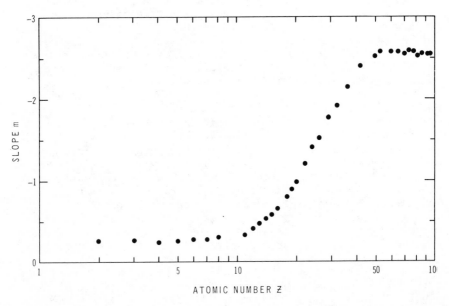

Figure 6.   Slope m of log μ(E) vs. log E between 100 and 150 keV vs. atomic number Z

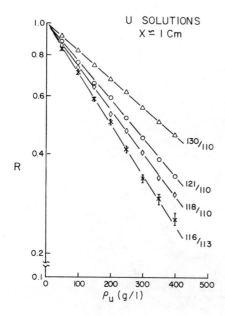

Figure 7. Ratio of transmissions, R, as a function of U concentration for 1-cm thick solutions and four transmission energy combinations

Figure 8. U solution assay system (USAS), a high-resolution gamma ray system for the measurement of U solutions at LASL's U recovery plant

assay the product solutions of a uranium recovery facility.
These solutions are well characterized and have been duplicated
by a set of standards with similar matrix properties. Routine
assays of 2-cm-thick samples for 10^3 s yield 1σ values of 0.5-1%
over a ρ_U range of 100-400 g/ℓ. Figure 9 summarizes the
measured uncertainties as a function of uranium concentration
for a fresh and a one-half-life-old transmission source. For a
fixed counting time, the uncertainties increase at lower concentrations due to the mean areal density being exceeded and at
higher concentration due to the poorer counting precision obtained for smaller transmission values. A discussion of
densitometry measurement design is given in ref. 17.

The reasonable penetrability of gamma rays near the uranium
and plutonium K-edges allows the densitometry technique to be
applied to some solids. For example, the ^{169}Yb transmission
source has been used to determine the number of plates in
materials test reactor (MTR) fuel bundles (18). Transmission
measurements were made at 109.8, 130.5, 177.2, and 198 keV for
1, 4, 18, and 19 MTR plates. To remove the effects of the
aluminum in the bundles, the measured transmission values, T(E),
were extrapolated to the K-edge of the uranium at 116 keV. The
extrapolation was performed by fitting $[\ell n(-\ell n\ T(E))]$ versus
$\ell n E$, where E is the gamma-ray energy, to a straight line. The
results of these measurements are shown in Fig. 10 where the
extrapolated ratio of transmission is plotted versus the number
of plates. To verify that a bundle contains the nominal 18
plates, it is necessary to measure the uranium in the bundle to
a one sigma accuracy of one quarter of a plate or 1.4%. This
requires that R be measured to within 3%, which is easily attainable. Because of the matrix insensitivity of the technique,
this verification procedure is not influenced by the substi-
tution of plates constructed from other materials.

An instrument is under development that adapts the K-edge
densitometry technique to the assay of cans containing plutonium-bearing ash (Fig. 11). The principal component of the
device is a drive assembly that translates the can in a
direction transverse to its cylindrical axis, between the
transmission source and the detector, while simultaneously
rotating it about its cylindrical axis. The transmission
sources, 50 mCi of ^{75}Se and 25 mCi of ^{57}Co, are mounted in a
large rotatable tungsten shield collimator. The total plutonium
mass in the can is given by

$$M_{Pu} = \frac{-\pi r^2}{\Delta\mu}\ \ell n\overline{R}$$

where $\overline{R}$ is the ratio of the gamma-ray transmissions averaged

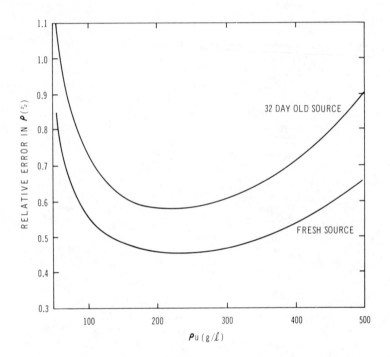

Figure 9. Measured uncertainties as a function of U concentration for a fresh and one-half life old ¹⁶⁹Yb transmission source

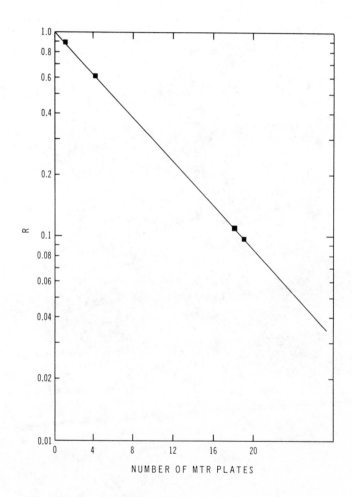

Figure 10. Ratio of transmissions, R, as a function of the number of plates for MTR fuel

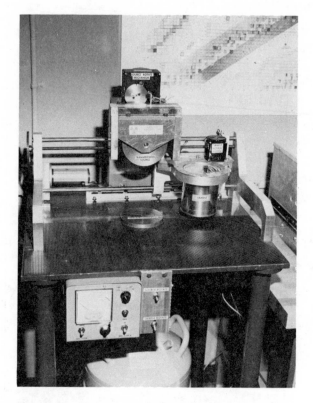

*Figure 11. Can scan, a high-resolution K-edge densi-
tometry device for the measurement of Pu-bearing ash
in cans*

over the can radius r. Preliminary results indicate that cans
with radii of ∿ 4 cm and containing 2 to 75 g of plutonium can
be assayed with this technique. Larger masses of plutonium can
be assayed in containers with correspondingly larger radii.

This discussion has been directed entirely towards use of
the K-edge signatures, as no obvious gamma-ray or x-ray emitting
radioisotopes are available that bracket the L_{III} edges.
However, one approach that may be applicable, similar to that
described in ref. 10, is to fluoresce a secondary target whose K
x-ray energies bracket the desired L_{III} edge. Low-energy
bremsstrahlung sources, which are discussed in the following
section, are, however, immediately applicable.

B.   Bremsstrahlung Sources

The energy spectrum from an x-ray generator may be tailored
by using filters and selecting the tube high voltage to yield
photons only in the immediate vicinity of the absorption edge of
interest. In addition, the photon rate in this optimized
spectrum is variable, allowing the detector count rate to be
maximized from sample to sample. The continuous nature of the
spectrum allows the influence of the matrix constituents to be
reduced to a minimum and the absolute sensitivity of the
technique to be maximized.

Figure 12 shows the spectra from a filtered 160-keV x-ray
beam transmitted through four 2-cm-thick plutonium-bearing
solutions and the resulting calibration curve obtained by taking
the ratio of counts on each side of the K-absorption edge at
121.8 keV. The ΔE from the edge is determined primarily by the
detector resolution, in this case 600 eV. To remove any matrix
sensitivity, the data on each side of the edge may be fit to a
simple function and extrapolated to the edge energy.

The higher sensitivity available at the L_{III} absorption
edges allows solutions containing lower concentrations of SNM,
$2 \lesssim \rho_s < 100$ g/ℓ, to be assayed accurately with manageable sample
thicknesses. Figure 13 shows the 23-keV x-ray spectra transmitted through a collection of uranium-bearing solution
standards (2 cm thick), as measured with a high-resolution
Si(Li) detector (19). The resolution in this case is approximately 250 eV.

Unlike assays at the K-edge, L_{III}-edge measurements are
sensitive to interfering K-absorption edges from low Z-matrix
constituents. The K edges of yttrium and zirconium, at 17.04
keV and 18.00 keV, are near the L_{III} edges of uranium and
plutonium respectively. The K edges of all other elements are
easily resolved from those of the SNM.

The continuous nature of the bremsstrahlung spectrum allows
multiple, simultaneous elemental concentration determinations.
Fig. 14 shows three spectra obtained with a filtered 150-keV

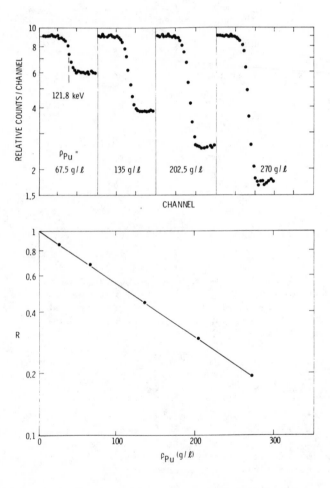

Figure 12. X-ray transmission spectra and the ratio of trans-missions for 2-cm thick Pu-bearing solutions (the K-edge of Pu at 121.8 keV is indicated)

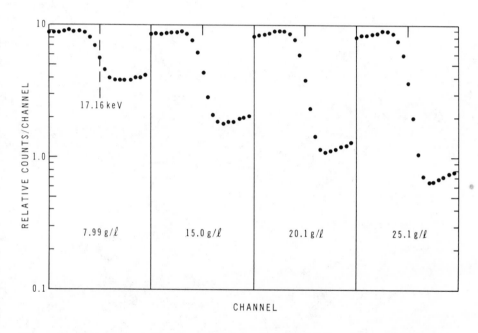

Figure 13. X-ray transmission spectra for 2-cm thick U bearing solutions about the L_{III} absorption edge of U (17.16 keV)

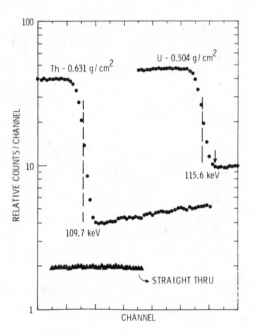

Figure 14. X-ray transmission spectra obtained with a filtered 150 keV x-ray generator. These spectra, one with no sample present, one transmitted through a Th foil, and one transmitted through a U foil, illustrate the potential for simultaneous determination of multiple elemental concentrations by the densitometry technique.

x-ray generator beam, one with no sample present, one transmitted through a thorium foil, and another transmitted through a
uranium foil. The channels corresponding to the thorium and
uranium edges are indicated. Figure 15 shows the spectrum
obtained when approximately 2 cm of HTGR fuel are substituted
for the foils. The K-absorption edges of both thorium and
uranium are clearly seen, allowing an accurate NDA determination
of the thorium-to-uranium density ratio.

This feature of absorption edge densitometry has direct
application in the reprocessing and coprocessing areas of the
nuclear fuel cycle. Figure 16 shows a filtered 23-keV x-ray
spectrum transmitted through a 1.1 cm thick solution containing
37 g/ℓ of uranium and 6.2 g/ℓ of zirconium. The discontinuity
at the zirconium K-edge (18.00 keV) approximates that which
would arise from the plutonium L_{III}-edge (18.07 keV) with 9
gPu/ℓ. This spectrum is representative of that which will be
obtained with an in-line densitometer that is presently being
developed for application to an experimental co-processing study.

Solutions are brought to the densitometer sample cell (Fig.
17) by lines extending from the rear of a glove box. For safety
considerations, the sample cell is enclosed in a secondary container, which is maintained at a negative pressure. The air
flow is past a solid state detector-based alpha alarm that will
alert personnel to any leaks within the secondary containment.
The system is designed to measure pure uranium streams with
concentrations of 20 to 70 gU/ℓ and mixed uranium-plutonium
streams with Pu/U ratios of 0.12-.25 (22-37 gU/ℓ, 2-6 gPu/ℓ).

Because of the spectrum complexity involved with the Pu-U
solutions, a different approach has been taken in the data reduction. In the absence of SNM in the sample, the transmitted
spectrum is smooth. The discontinuities introduced into the
spectrum by SNM absorption edges, convoluted with the detector
resolution function, appear as peaks in the derivative function. An example is shown in Fig. 18 where a channel by channel
difference of the natural log of the data shown in Fig. 16 is
presented. The net areas of the U and Zr peaks, which may be
determined by standard fitting routines, is equal to lnR.

III.   FUTURE APPLICATIONS
The examples of the densitometry technique discussed in the
previous section clearly illustrate that the method is versatile
and relatively easy to implement. In the future, densitometry
should see considerable usage in three major areas of application of nuclear assays: the analytical laboratory, in-line
instrumentation for safeguards and process control, and portable
instrumentation for use by international or regulatory
inspectors.

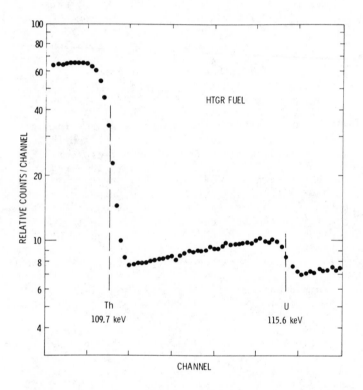

Figure 15. X-ray transmission spectrum for HTGR fuel. The K-edge of Th at 109.7 keV and U at 115.6 keV are shown.

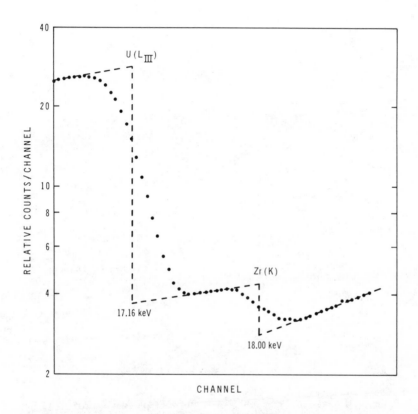

Figure 16. X-ray transmission spectrum for a mixed uranium–Zr solution. Superimposed is an infinite resolution spectrum (dashed lines) showing the position of the U L_{III}-edge at 17.16 keV and the Zr K-edge at 18.00 keV.

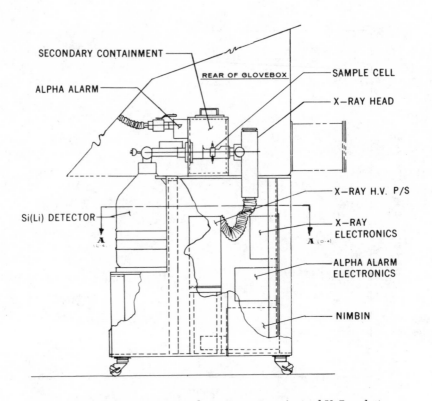

Figure 17. In-line densitometer for the measurement of mixed U–Pu solutions. This device uses the L_{III} edges of U and Pu to perform this measurement.

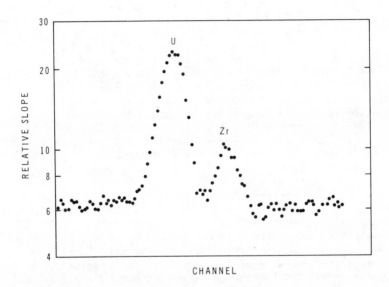

Figure 18. Simple derivative filter applied to the x-ray transmission spectrum in Figure 16. The peaks are the U and Zr edge information. The area under the peaks is proportional to the elemental concentration. The width is a function of detector resolution.

A. The Analytical Laboratory

The analytical laboratory associated with the large nuclear fuel cycle facilities of the future will require assay techniques which are rapid, relatively inexpensive and easy to perform reliably on a routine basis. A densitometer consisting of a x-ray generator capable of producing photon beams with energies from 15 to 150 keV, a planar Ge detector with a resolution at 122 keV of 500 eV or better and associated electronics, and a computer based data acquisition and reduction system can be otained for a capital expenditure of approximately $80,000. Such a device could assay samples of U, Pu, Th or mixtures thereof with concentration ranging from a few up to hundreds of g/ℓ by utilizing both the L_{III} and K-edge technique. Minimal sample preparation can extend the range of the method. Properly programmed for the routine problems expected in the facility, the densitometer system could be operated by a relatively untrained technician. Assay times and precisions to be expected with such a device will depend in detail on the material concentrations and matrices, but experience to date has shown that assays better than 0.5% RSD in 1000 s assay times are obtainable for pure (elemental) SNM materials in solutions with concentration greater than 10 g/ℓ.

B. In-line Instrumentation

Real time or near real time knowledge of SNM concentrations in the flowing streams of a particular process is essential for the application of modern safeguards accounting systems and process control. The co-processing densitometer (described above) is a good example of the use of an in-line instrument with a laboratory scale reprocessing facility. Larger plant-scale operations will be able to utilize this design on measuring by-pass streams drawn from the major flow.

For applications where only single SNM elements will be present in lower z matrix solvents, it is attractive to contemplate the use of simpler (and less costly) in-line instruments with precisions and accuracies which are larger than those required in the analytical laboratory. Such instruments would employ poorer resolution detectors operating at room temperature and less versatile electronics packages which might provide only a simple readout of SNM concentration. Transmission sources would be either radioisotopic gamma-ray or bremsstrahlung sources. Such densitometers would be sacrificing measurement precision and accuracy in favor of cost, thus allowing the placement of more instruments. · A system employing a NaI(Tℓ) detector and electronics for K-edge assays might cost $10-15,000.

C. Portable Instrumentation

Safeguards inspectors, both international and domestic, are frequently faced with the problem of accounting for SNM in a

variety of forms using only portable assay equipment. A knowledge of densitometry techniques and some specialized equipment
for densitometry measurements would be a valuable addition to
the inspectors repertoire. The example of verification of MTR
fuel bundles mentioned in Section II could be implemented in the
field using portable Ge detectors and compact multichannel
analyzers already being used by the IAEA. Less precise instrumentation of the type already described for in-line application
could also be developed as a dedicated, portable densitometer
for liquid samples. Further developments in the area of good
resolution, compact gamma- and x-ray detectors such as CdTe and
HgI_2 will have a considerable impact on the design of portable
instrumentation and would be especially important for
densitometry.

Finally, it is important to note that densitometry, when
combined with other analytical tools and techniques, can provide
information which is otherwise very difficult to obtain. As
examples, consider the assay of SNM contained in highly radioactive solutions and uranium enrichment measurements in
solutions. A Bragg curved crystal spectometer (with "poor"
resolution) set to pass only that portion of the transmission
beam surrounding the L_{III} edge of interest can be used as a
narrow band-pass filter which allows the densitometry
measurement to be performed, but blocks the high background due
to the sample (ref. 20). Uranium enrichment measurements in
solutions (or other low density media) require an assay of both
^{235}U and total U. The concentration of ^{235}U can be determined from transmission corrected passive gamma-ray counting
while total U concentrations are given by the densitometer.

122   NUCLEAR SAFEGUARDS ANALYSIS

REFERENCES

868888 8
1. R. E. Barieau, "X-Ray Absorption Edge Spectrometry as an Analytical Tool - Determination of Molybdenum and Zinc," Anal. Chem. 29, 348-352 (1957).

2. H. W. Dunn, "X-Ray Absorption Edge Analysis," Talanta 6, 42-45 (1960).

3. E. A. Hakkila and G. R. Waterbury, "X-Ray Absorption Edge Determination of Cobalt in Complex Mixtures," Advan. X-Rays Anal. 5, 379-388 (1961).

4. E. A. Hakkila and G. R. Waterbury, "Applications of X-Ray Absorption Edge Analysis," Devel. Appl. Spect. 2, 297-307 (1963).

5. C. G. Dodd and D. J. Kaup, "Optimum Conditions for Chemical Analysis by X-Ray Absorption-Edge Analysis," Anal. Chem. 36, 2325-2329 (1964).

6. K. T. Knapp, R. H. Lindahl, and A. J. Mabis, "An X-Ray Absorption Method for Elemental Analysis," Advan. X-Ray Anal. 7, 318-324 (1964).

7. E. P. Bertin, R. J. Longoloucco, and R. J. Carver, "A Simplified Routine Method for X-Ray Absorption Edge Spectrometric Analysis," Anal. Chem. 36, 641-655 (1964).

8. H. W. Dunn, "X-Ray Absorption Edge Analysis," Anal. Chem. 34, 116-121 (1961).

9. W. B. Wright and R. E. Barringer, "Uranium Analysis by Monochromatic X-Ray Absorption," Union Carbide Nuclear Company Report Y-1095 (August 1955).

10. E. A. Hakkila, "X-Ray Absorption Edge Determination of Uranium in Complex Mixtures," Anal. Chem. 33, 1012-1015 (1961).

11. E. A. Hakkila, R. G. Hurley, and G. R. Waterbury, "Three-Wavelength X-Ray Absorption Edge Method for Determination of Plutonium in Nitrate Media," Anal. Chem. 38, 425-427 (1966).

12. T. R. Canada, J. L. Parker, and T. D. Reilly, "Total Plutonium and Uranium Determination by Gamma-Ray Densitometry," ANS Transactions 22, 140 (1975).

13. T. R. Canada, J. L. Parker, and T. D. Reilly, "Total Plutonium and Uranium Determination by Gamma-Ray Densitometry," LA-6040-PR, 9-12 (1975).

14. T. R. Canada, R. C. Bearse, and J. W. Tape, "An Accurate Determination of the Plutonium K-Absorption Edge Energy Using Gamma-Ray Attenuation," NIM 142, 609-611 (1977).

15. T. R. Canada, D. G. Langner, J. L. Parker, and E. A. Hakkila, "Gamma- and X-Ray Techniques for the Nondestructive Assay of Special Nuclear Material in Solution," LA-6881, Vol. II, 1-38 (1977).

16. R. Gunnink and J. E. Evans, "In-Line Measurement of Total and Isotopic Plutonium Concentrations by Gamma-Ray Spectrometry," UCRL-52220 (1977).

17. J. W. Tape, D. G. Langner, J. L. Parker, and T. R. Canada, "The Measurement of Special Nuclear Material Concentrations in Solution by Absorption Edge Densitometry in Analytical Chemistry in Nuclear Fuel Reprocessing," W. S. Lyon (Ed.), Science Press (1978).

18. D. G. Langner, S. T. Hsue, and T. R. Canada, "Determination of the Total Uranium Content in Materials Test Reactor (MTR) Fuel by K-Edge Densitometry," LA-6788-PR, 27-29 (1977).

19 T. R. Canada, S. T. Hsue, D. G. Langner, E. R. Martin, J. L. Parker, T. D. Reilly, and J. W. Tape, "Applications of Absorption-Edge Densitometry NDA Technique to Solutions and Solids," Journal of INMM 6, 702-710 (1977).

20. J. W. Tape, T. R. Canada, and E. A. Hakkila, "Application of Dispersive X-Ray Filtering Techniques to Absorption-Edge Densitometry," LA-6849-PR, 10-11 (1977).

RECEIVED JUNE 20, 1978.

9

Application of On-Line Alpha Monitors to Process Streams in a Nuclear Fuel Reprocessing Plant

K. J. HOFSTETTER, G. M. TUCKER, R. P. KEMMERLIN, J. H. GRAY, and G. A. HUFF

Allied-General Nuclear Services, Barnwell, SC 29812

The detection of alpha activity using scintillating glasses was first developed by Upson(1) and later applied to plant process streams by Huck and Lodge.(2) The sensor chosen for plant application was cerium-activated Vycor glass developed by Corning Glass Works in cooperation with Hanford Laboratories. Improvement of the sensors' ability to detect low levels of alpha activity in streams containing high levels of beta activity was suggested by Koski.(3) By careful control of the cerium-activated zone geometry, Koski found that the beta sensitivity could be minimized without sacrifice of the alpha efficiency. As alpha monitoring of plant process streams could provide valuable data in operating a nuclear fuels reprocessing plant, a contract was undertaken with Intelcom Rad Tech (IRT) to develop such a series of monitors.

Gozani, et al.(4) at IRT employed the cerium-activated Vycor sensors in the design of the On-Line Alpha Monitors for use on process streams at the Barnwell Nuclear Fuel Plant (BNFP). The lines to be monitored are listed in Table 1. Each sensor was optimized to that particular sample stream to maximize the alpha sensitivity while minimizing the beta sensitivity. The liquid sampling cell included a deflector baffle to constrict and direct the flow of liquid across the face of the detector. The thickness of the solution sampled varied between 0.005 and 0.020 inches depending on the expected activities in the sample streams. All internal surfaces which come into contact with radioactive solution were highly polished to minimize holdup of the process solution.

In the IRT development study, the OLAMs were tested by exposing the monitors to alpha and beta radioactive point sources and encapsulated liquid sources. Pulse height distributions and discriminator curves were obtained using these sources for each monitor and supplied with the OLAMs.

The OLAMs have been subjected to a testing program at BNFP to evaluate detector performance when exposed to acid solutions containing plutonium. A laboratory system was designed to circu-

TABLE I

TYPICAL STREAMS AT BNFP MONITORED WITH OLAMS

Monitor Point	Concentration g Pu/ℓ	β, γ Activity dps/ml	
1CU Stream	2×10^{-4}	2.6×10^{5}	aq; 0.17$\underline{M}$ HNO$_3$, 0.28$\underline{M}$ U
1BP Stream	4.82	1.7×10^{6}	aq; 2.9$\underline{M}$ HNO$_3$, 0.04$\underline{M}$ U
2AW Stream	4×10^{-3}	1.1×10^{6}	aq; 2.85$\underline{M}$ HNO$_3$
2BW Stream	1.4×10^{-2}	1.7×10^{2}	org; 0.04$\underline{M}$ HNO$_3$, 0.08$\underline{M}$ U
1SW Stream	2×10^{-4}	2.85×10^{6}	aq; 2.2$\underline{M}$ HNO$_3$
1SF Stream	5×10^{-3}	2.85×10^{6}	aq; 2.2$\underline{M}$ HNO$_3$, 0.01$\underline{M}$ U
3AW Tank	1.5×10^{-2}	1.08×10^{3}	aq; 2.85$\underline{M}$ HNO$_3$
3BW Stream	1×10^{-3}	low	org; 0.09$\underline{M}$ HNO$_3$, 0.04$\underline{M}$ U
POR Stream	1.7×10^{-2}	95	org; 0.06$\underline{M}$ HNO$_3$, 0.06$\underline{M}$ U
3PD KO Pot	7×10^{-3}	low	aq; 0.33$\underline{M}$ HNO$_3$

late solutions through an OLAM with the capability for sampling
test solutions. The response of the OLAM to plutonium solutions
in the concentration range of 10^{-4} to 3 g/ℓ established the sensitivity, accuracy, reliability, long term stability and useful
concentration range for streams containing typical low burn-up
light water reactor grade plutonium. The ability of the monitor
to discriminate between alpha and beta activity was determined by
adding ^{90}Sr to the plutonium solutions. Other interferences were
studied along with detector decontamination procedures.

EXPERIMENTAL

The detector chosen for study as typical of the OLAMs used
at BNFP was the spare system for the 1BP Stream. The 3-inch
diameter sensor was coupled to a 2-inch diamter RCA-4507 photo-
multiplier tube. The cerium-activated layer of the detector
chosen for study is about 1.2 inches in diameter. The clearance
between the detector and the liquid flow baffle is 0.011 inch.
The electronics required to operate the OLAM include a voltage
divider, a signal decoupling unit, (as both high voltage and
signal use the same coaxial cable) an amplifier, high voltage
supply, discriminator and ratemeter. In addition to ratemeter-
discriminator studies, the amplifier output was interfaced to a
computer-based multichannel analyzer for pulse height analysis.
Spectra could be recorded for preset time, spectral analysis
performed, and the results stored on disc or magnetic tape.

A point source holder was designed to accomodate the OLAM
head assembly. Using this device, spectra of radioactive point
sources could be taken without interference from light leaks,
and the sources could be changed without turning off the high
voltage.

Spectra were recorded using ^{241}Am, ^{238}Pu and ^{148}Gd NBS-cali-
brated alpha-particle sources. A spectrum of ^{238}Pu taken with
the OLAM is shown in Figure 1. The full-width at half-maximum
resolution of the alpha-peak was determined to be 30%.

Sequential spectra of the ^{238}Pu point source were recorded
for 10 second intervals intermittently for several days. The
resultant peak areas are shown in Figure 2. A program was written
to acquire a spectrum for a preset time, integrate the alpha-peak,
store the results, clear the data region and return to the acquire
mode. After a preset number of cycles, the average area and associated error were printed out. Each data point in Figure 2
represents the average area for 200 spectra. The relative standard deviation for these data is 1.5% (one sigma). These data
indicate good short-term stability of the detector and electronics
and fair source reproducibility.

After preliminary testing of the OLAM in the laboratory, a
solution circulation system located in a glove box was fabricated
so that plutonium-bearing solutions would flow through the monitor.
A sampling port was included. A 2.6M HNO$_3$ solution was used to
simulate the solution concentration seen by a typical OLAM during
plant operation. Preliminary experiments with the OLAM indicated

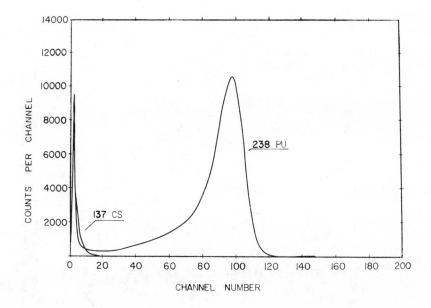

Figure 1. Spectra of ^{238}Pu and ^{137}Cs point sources recorded with the on-line alpha monitor used in this study

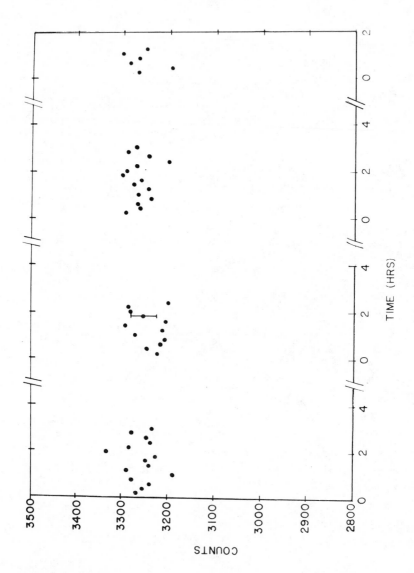

Figure 2. Total alpha counts per 10-sec interval recorded over a period of several days. Each point represents the average of spectra with a typical error bar shown.

the maximum solution flow rate through the detector was 1000 ml/min, well above the anticipated maximum in-line sampler flow rate of 250 ml/min. Standard 1/4 inch stainless steel tubing was used in the fabrication of the recirculation system flow lines. The system is shown pictorially in Figure 3.

A bellows pump P is used to pump liquid from the reservoir B to reservoir A. The liquid is then allowed to flow by gravity through the monitor which is mounted at 45° to minimize bubbles sticking to the detector surface. The flow through the OLAM is regulated by valve S and monitored by flow meter F. Sections of Tygon tubing were used to introduce some flexibility into the system. The maximum flow rate through the test system was found to be 125 ml/min. A dual reservoir system was constructed after observation of "pulsing" of the liquid through the monitor when using a closed system. Vent ports V1 and V2 in the two reservoirs eliminated the pulsing effect.

Liquid in the system was drained through valve D. Vent port V1 was used to withdraw 100 λ samples for plutonium concentration measurements without disturbing the system. The volume of solution in the system is 125 ml. With both reservoirs open to the atmosphere in the glove box, the solution evaporated slowly thus increasing the plutonium concentration. The extent of the evaporation can be seen in Figure 4, where the average number of counts for 500 spectra taken for 30 seconds each are plotted as a function of time. The sawtooth curve shows the response of the OLAM to the concentration increase with evaporation, followed by subsequent dilution. About 30 ml of the solution typically evaporated in one week. The solution was allowed to concentrate for one week (about 20,000 spectra recorded during this time) then diluted with 2.6M HNO_3. The temperature of the system affected the rate of solution evaporation. This can be seen by noting the change in slope of each cycle on the plot in Figure 4. The ratemeter also reflected the continuous increase in plutonium concentration during a one week cycle.

Plutonium nitrate solutions were prepared in 2.6M HNO_3 at different levels of plutonium concentrations. These solutions (10^{-4} to 3 g/ℓ) were introduced into the system and studied for 3-4 weeks to obtain long-term stability data on the monitor. Samples of the solution were taken daily and analyzed for plutonium content by alpha counting. The isotopic composition and the alpha-specific activity of the plutonium stock solution are given in Table II along with compositions of typical light water reactor grade plutonium (see later discussion).

Data for count rate as a function of plutonium concentration were taken by recording the OLAM spectra in a time differential mode with the computer-based multichannel analyzer. Computer programs were written to collect and analyze the data simultaneously. For each concentration range of plutonium studied, a minimum of 80,000 spectra were recorded and analyzed. At the termination of the experiment, the plutonium solution was drained by opening

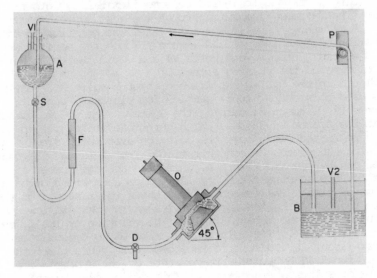

Figure 3. Schematic of the solution circulation system used to test the on-line alpha monitors.

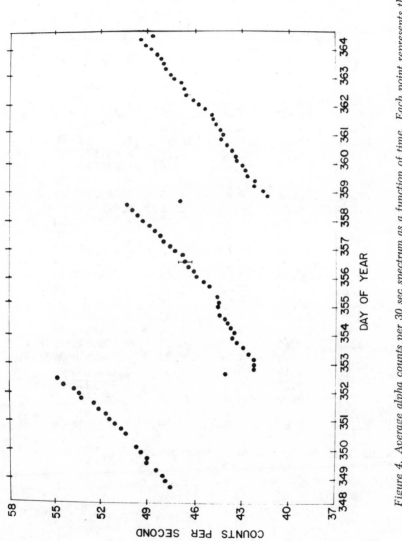

Figure 4. Average alpha counts per 30 sec spectrum as a function of time. Each point represents the average of 500 spectra.

TABLE II.

ISOTOPIC COMPOSITIONS AND SPECIFIC ACTIVITIES
OF PU IN TYPICAL PWR FUEL

Plutonium Isotopic Composition					Specific Activity
238	239	240	241	242	ALPHA DPS/GM
1.92	63.30	19.20	11.70	3.88	1.523290E+10
0.42	75.40	15.60	7.33	2.10	5.709610E+09
1.71	53.30	28.80	10.30	5.85	1.448650E+10
0.57	66.40	23.40	7.30	2.35	7.110770E+09
1.55	45.90	30.10	11.40	11.10	1.342310E+10
0.14	87.00	9.56	3.06	0.21	3.687570E+09
0.34	80.60	12.70	5.69	0.69	5.074370E+09
0.56	76.30	14.50	7.49	1.13	6.521920E+09
0.65	72.90	16.50	8.41	1.55	7.183630E+09
1.00	67.70	18.80	10.00	2.51	9.475780E+09
1.92	63.30	19.20	11.70	3.88	1.523290E+10
1.93	63.20	19.30	11.80	3.73	1.530220E+10
2.15	56.40	21.90	13.80	5.77	1.676240E+10
0.42	75.40	15.60	7.33	1.22	5.708340E+09
0.52	74.10	16.00	7.98	1.42	6.345880E+09
0.60	73.20	16.20	8.48	1.56	6.848950E+09
0.75	71.20	16.90	9.34	1.85	7.812420E+09
0.78	70.80	17.00	9.43	1.93	8.001690E+09
0.83	71.00	17.00	9.19	1.94	8.322400E+09
0.03	84.80	12.50	2.50	0.13	3.188600E+09
0.08	78.70	16.30	4.56	0.33	3.688000E+09
0.08	77.10	17.10	5.34	0.42	3.719700E+09
0.10	74.40	18.90	6.08	0.55	3.937120E+09
0.10	73.70	19.30	6.32	0.59	3.955110E+09
* 0.01	82.19	16.30	1.24	0.26	3.362260E+09

* Plutonium used in the course of this study.

stopcock D. About 15% of the plutonium remained in the monitor in traces of residual solution or fixed on the walls of the system. A flush solution containing 125 ml of 2.6$\underline{M}$ HNO$_3$ was added to the system and allowed to circulate for a minimum of 24 hours during which approximately 3,000 more spectra were recorded and analyzed. The following day, the flush solution was drained after sampling and another 125-ml portion of fresh 2.6$\underline{M}$ HNO$_3$ was added. The flushing process was continued until the concentration of plutonium was less than 1x10^{-5} g/ℓ. In general, five flushes were required. Typical 22-hour accumulation spectra taken with the monitor are shown in Figure 5 for a sequential series of flushing runs.

It is apparent from Figure 5 that the character of the spectrum changes as the concentration of the plutonium is diluted. The fourth flush spectrum resembles the spectrum of the point source shown in Figure 1 as a definite alpha-peak is present. These data indicate that the plutonium is "plating out" on the cerium-activated Vycor detector surface. For solution streams containing low plutonium concentrations, a significant correction will need to be applied in monitoring the alpha-activity.

The experiments designed to measure the extent of beta interference in the alpha spectrum were performed using ^{90}Sr as a spike. A stock solution (0.4 mCi ^{90}Sr/ml) was prepared. After circulation of a plutonium solution for several days, an aliquot of the ^{90}Sr stock solution was introduced into the system. Discriminator curves using the discriminator and rate meter along with spectra were recorded of the OLAM output. Samples were taken and analyzed for alpha and beta activity. After several days of circulation another ^{90}Sr spike was added and the sequence repeated until the upper limit in beta activity was reached.

The ratio of alpha to beta activity for the pure plutonium solution was about 50. As more ^{90}Sr spike was added, the ratio of alpha to beta activity decreased until a minimum value of 0.2 was reached. The specific beta activity of the solutions tested including flush out spectra ranged from 10^3 to 10^8 dpm/ml which simulates plant conditions. Similar experiments will be performed using ^{137}Cs beta-gamma activity.

RESULTS

The short-term count rate stability of the OLAM system has been evaluated using point sources and found to be better than 1% relative standard deviation (RSD). The peak centroid stability was found to be better than 2% RSD. These data indicate stable operation of the detector-amplifier system.

The plutonium isotopic composition used in this study simulates only low burn-up light water reactor fuel. Table II gives the plutonium isotopic abundances of light water fuel from 24 reactor cores of the pressurized light water reactor type. The isotopic composition of the plutonium used in this study is given for reference. The reactor data are taken from Smith, et al. (5) It can be seen that the plutonium used in this study represents

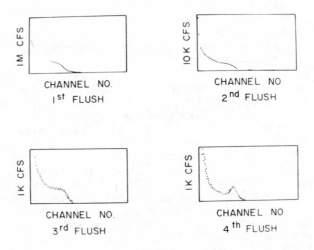

Figure 5. Typical spectra taken during a series of flush runs. Each spectrum was taken for ~ 22 hr (CFS means counts full scale).

very low burn-up fuel as compared to the specific activity of high
burn-up fuel (e.g., fuel number 1 having 1.92% ^{238}Pu). The range
of specific activities can be seen to vary a factor of 6 for the
data presented. An accurate determination of the specific alpha-
activity is required to compute the plutonium concentrations from
alpha count rate data.

For the plutonium used in this study, Figure 6 shows how the
count rate changes with concentration. On a log-log plot many of
the variations have been dampened. A linear relationship is suggested by this figure over a large range of plutonium concentrations. Each data point represents the count rate average of 500
spectra taken during the course of the sampling. The data points
cluster about the prime plutonium concentrations with the intermediate data points coming from the solution flushes.

To more accurately display the deviations at high and low
plutonium concentration, the inverse efficiency was computed for
each sample by dividing the solution alpha-specific activity
(dpm/ml) by the monitor count rate (cps). This ratio should be
independent of plutonium concentration. Figure 7 is a plot of
the deviations of the experimental values from the least squares
value of the ratio computed at the medium concentrations. For
plutonium concentrations between 10^{-3} and 10^{-1} g/ℓ, the one sigma
relative standard deviation fit to the data is 11.8%. The data
at the concentration extremes exhibit significant negative deviations. The deviations at low plutonium concentrations are primarily due to the plating out effect giving an apparent increase
in sensitivity.

An apparent increase in sensitivity is also observed at high
plutonium concentrations. During these experiments, the lower
level discriminator was set just above the system noise level.
Both the alpha and beta portion of the plutonium spectrum were
recorded yielding a total system count rate of 40,000 cps. At
these count rates, dead time losses and pulse pile-up could cause
some deviations. For a typical run to generate one data point,
the results for 500 spectra were averaged. A large RSD (3%), was
observed at 3 g/ℓ plutonium solution when compared to a typical
run at 10^{-2} g/ℓ (0.7% RSD). As the beta response of the OLAM is
exponential(3), any discriminator instability will be amplified
at low levels.

The beta interference studies were conducted using ^{90}Sr
spiked into plutonium solutions. Spectra were recorded and
samples taken and analyzed for alpha and beta activity. Discriminator curves were also taken using the rate meter and discriminator. A typical spectrum is shown in Figure 8 for a solution of
0.01 g/ℓ plutonium with beta activity levels from 10^3 to 10^8 dpm
^{90}Sr/ml. The discriminator setting required to eliminate the
beta-activity portion of the spectrum is 30% of the alpha full
energy peak. With the discriminator set at this level, the sensitivity was determined to be 52.3 ±1.9 cps/μCi/ml and independent
of the beta-activity level. A plot of the percent deviations as

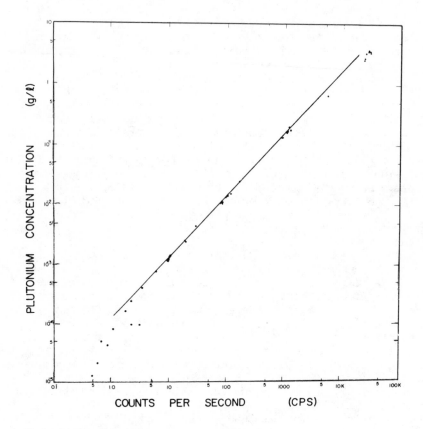

Figure 6. Plot of the OLAM count rate in counts per sec as a function of Pu
concentration in g/L

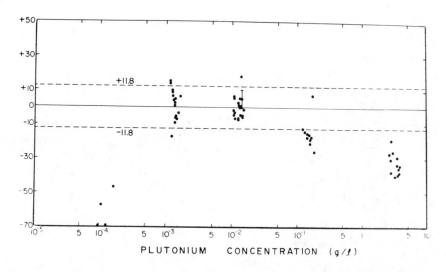

Figure 7. *Plot of the deviations of the data shown in Figure 6 from a linear response function*

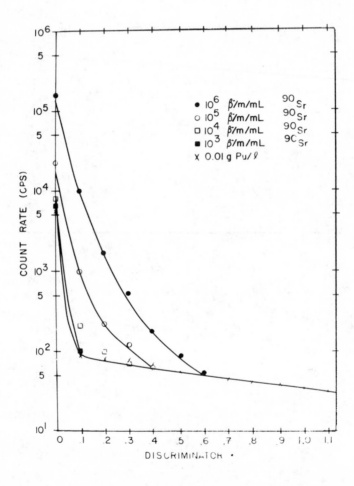

Figure 8. Spectra of Pu solutions that have been spiked with various quantities of ^{90}Sr showing the extent of beta spectral interference

a function of beta-activity is shown in Figure 9 at 0.01 g/ℓ.
With a 30% discriminator, an experiment at 0.06 g/ℓ yielded a
sensitivity of 52.8 ±2.0 cps/μCi/ml. The deviations as a function
of beta concentration are shown in Figure 10. With 1.4 g/ℓ plutonium solution, a sensitivity of 46.7 ±1.9 cps/μCi/ml was observed. These data are shown in Figure 11.

It is apparent that for the monitor tested, a 30% discriminator setting is sufficient to eliminate most levels of beta interference expected on the plant process streams. As the thickness of each sensor is different, the discriminator level for
each monitor will need to be determined for optimum performance.
The alpha and beta responses can be determined using point
sources, once the general operational characteristics have
been identified. There is a decrease in alpha-senstivity with an
increase in discriminator level. With no discrimination, the
sensitivity was found to be 80.8 ±9.5 cps/μCi/ml assuming that all
counts were alpha events. It is apparent that some of those
counts were due to beta events. Experiments were performed at
varying plutonium concentrations using the OLAM with a discriminator setting of 30%. The deviations of the data from an average
sensitivity as a function of plutonium concentration are shown in
Figure 12. Each data point is the average of 500 spectra correlated with sampling data over the range of 10^{-4} to 2 g/ℓ plutonium
concentration. There is still evidence of a non-statistical trend
at the concentration extremes but the absolute deviation is much
smaller. The Alpha sensitivity computed with these data is
52.0 ±2.6 cps/μCi/ml.

During the high concentration runs, there was evidence of a
gain shift in the spectrum (about 10%). When the discriminator
is set just above the noise level, a small change in gain will
significantly affect the system count rate. At high count rates,
the relative standard deviation indicated non-statistical behavior
but after completion of the runs, the detector settled back to
normal behavior during the flush out experiments.

The amount of plutonium that plated out on the cerium-activated
Vycor detector was about 5-6 times the background rate observed in
the initial experiment where only 2.6M HNO₃ was circulated. No
special decontamination agents have been tested to date. Later
experiments will be developed to decontaminate the sensor.

CONCLUSIONS

Based on the test data compiled in this study, the OLAMs to
be used at BNFP are stable devices that can determine the alpha-
activity in on-line process streams. The useful detection range
of the alpha monitors in the 2.6M HNO₃ is from 20,000 dpm/ml to
about 10^8 dpm/ml for solutions containing beta-activity from 10^3
to 10^8 dpm/ml at a flow rate of 125 ml/min. From the evaporation
curve data, the OLAMs respond to small changes in the concentration
(3% per day) with a high degree of accuracy. The response time
for sensing a rapid change in alpha-solution content is almost
immediate. At low concentrations, the plating out effect will

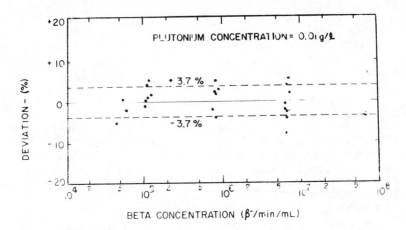

Figure 9. Percent deviations of the data from a constant response as a
function of ^{90}Sr concentration at 0.01 g Pu per L concentration

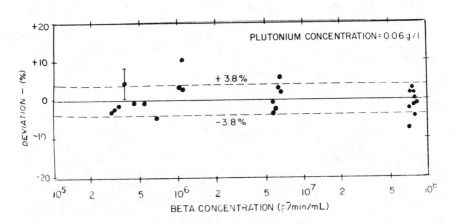

Figure 10. Deviations of the data from a linear constant as a function of ^{90}Sr concentration at a Pu concentration of 0.06 g/L

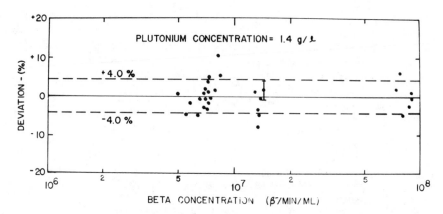

Figure 11. *Deviations of the data from a linear constant as a function of ^{90}Sr concentration at a Pu concentration of 1.4 g/L*

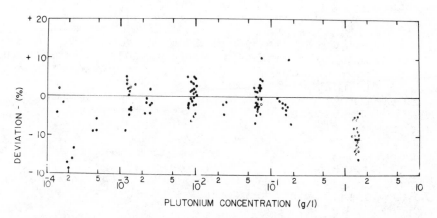

Figure 12. *Deviations of the data from a linear response as a function of Pu concentration using a 30% discriminator*

seriously affect the accuracy of the OLAM. A suitable acclimation time will be required at each concentration. At high alpha-activity levels, the OLAM begins to fail due to pulse pair resolution and pile-up effects, dead time losses, and gain changes.

In the next phase of this study, the effect of several parameters will be investigated. The effect of flow rate on the OLAM count rate will be determined by measuring the count rate at several different flow settings. The effect of gamma-activity will also be investigated. Plutonium solutions (in 2.6M HNO₃) of various concentration will be circulated and ^{137}Cs (beta, gamma emitter) spiked into the system. The gamma discrimination will be tested and sensitivity and accuracy of the OLAM redetermined in the presence of various levels of gamma interferences. The effect of uranium in the solutions will be measured as the bulk density of the solution will change along with the specific alpha activity. Later phases of this project will include duplication of the tests for organic streams, the calibration of each monitor, and the evaluation of pile-up rejection electronics to overcome some of the problems observed at high count rates.

ABSTRACT

The On-Line Alpha Monitors (manufactured by Intelcom Rad Tech and patterned after the design used at the Hanford Purex Facility) (OLAM) have been subjected to a testing program to evaluate detector performance when exposed to acid solutions containing plutonium. The OLAMs are shown to be stable devices that can measure alpha activity in process streams containing fission products. A laboratory system was designed to circulate solutions through the OLAM with a capability for sampling test solutions. The response of the OLAM to plutonium solutions in the concentration range of 10^{-4} to 3 g/ℓ established the sensitivity, accuracy, reliability, long term stability, and useful concentration range for typical low burn-up light water reactor grade plutonium with specific alpha activity of 2.0×10^{11} dpm/gm. A linear detector response to plutonium concentration was observed from 10^4 to 10^8 dpm/ml in 2.6M HNO₃ solutions. Plutonium concentration of these test solutions was determined by alpha spectrometry. The ability of the monitor to discriminate between alpha and beta activity has been determined by adding ^{90}Sr to the plutonium solutions. A discrimination factor of 10^4 beta events per alpha event can be obtained with the OLAM. Interferences from uranium in the solutions will be discussed along with detector decontamination procedures.

Literature Cited

1 Upson, U. L., "Scintillation Glasses for Alpha Counting", HW-72512, (1962).
2 Huck, C. E, and Lodge, J. D., "In-Line Plutonium Analyses by the Use of Alpha Scintillating Glass and X-Ray Scintillating Techniques", HW-84251, (1964).
3 Koski, O. H., "The Optimization of Cerium-Activated Glass Alpha Sensors", BNWL-CC-2354, (1969).

4   Gozani, T., and MacKenzie, J., "On-Line Alpha Monitor System", INTEL-RT-5011-002, (1974).
5   Smith, R. C., Faust, L. G., and Brackenbush, L. W., "Plutonium Fuel Technology Part II.   Radiation Exposure from Plutonium in LWR Fuel Manufacture," Nuclear Technology 18, 97-108 (1973).

RECEIVED MAY 24, 1978.

10

Uranium and Plutonium Analyses with Well-Type Ge(Li) Detectors[1]

F. P. BRAUER, W. A. MITZLAFF, and J. E. FAGER

Battelle Memorial Institute, Pacific Northwest Laboratories, P.O. Box 999, Richland, WA 99352

Analysis of microgram and submicrogram quantities of ^{235}U and ^{239}Pu are required by the nuclear industry for process control, nuclear safeguards and effluent measurements. These analyses are of increasing importance in efforts to reduce inventory discrepancies and uncertainties. Current analytical laboratory methods used for measurement of small quantities of uranium and plutonium include x-ray fluorescence methods, spectrophotometric methods, fluorometric methods, radiometric methods, and mass spectrometric methods (1,2,3). Many of these analytical laboratory methods measure only total plutonium and uranium while newer nondestructive analysis (NDA) methods, which have been developed primarily for in-plant use, can measure specific isotopes of uranium or plutonium (2,3,4). Adaptation of some of the NDA techniques to the analytical laboratory would result in more rapid and more specific analyses. This paper discusses an NDA method for rapid laboratory analysis of ^{239}Pu and ^{235}U.

Gamma-ray spectrometric methods can be used in the analytical laboratory for both direct measurement of sample aliquots (NDA) and for performing measurements on samples following laboratory processing. Samples can often be prepared for gamma-ray spectrometric measurements with considerably less effort than is required for measurement by other methods. Gamma-ray spectrometric methods can measure specific radionuclides, an important consideration in facilities processing enriched uranium. Gamma-ray spectrometric methods also differentiate between ^{241}Am and plutonium and can be used for plutonium isotopic analyses.

A well-type Ge(Li) detector was used for measurements on standard uranium ore, uranium and plutonium samples. This paper discusses the results of these measurements and the application of x-ray and gamma-ray spectrometric measurements to laboratory uranium and plutonium determinations.

Characteristics of the Ge(Li) Detector and Electronic Instrumentation

A commercial well-type Ge(Li) detector was used in these experiments. The detector has a nominal active volume of 70 cm^3, a diameter of 5.25 cm and a length of 4.40 cm. The detector's well accommodates samples up to 1 cm in diameter and 4 cm in length. The detector resolution was measured as a function of energy (Figure 1) and was found to be as good as similar sized non-well Ge(Li) detectors. The ^{137}Cs 662 keV photopeak-to-Compton edge ratio for a sample in the well of the detector was found to be 102.4.

A mixed radionuclide gamma-ray solution standard obtained from the National Bureau of Standards (SRM-4254) and an ^{241}Am solution standard were used to prepare standard sources for detector efficiency measurements. The detector efficiency (counts/gamma-ray) for the source in the well of the detector was measured as a function of energy (Figure 2). The detector was found to have high resolution and efficiency in the regions of interest for ^{235}U and plutonium gamma-ray and x-ray spectrometric measurements.

Background measurements were made with the detector in a lead shield 10 cm thick. The background was found to be similar to that observed with Ge(Li) detectors of similar dimension without wells. The major background photopeaks are from lead x-rays, which can be reduced by shield design, and other peaks from natural radioactivity. Background reductions obtained with an anticoincidence shield have been previously reported (5).

The electronic system used for data acquisition and analysis includes a spectroscopy amplifier, a nuclear 8192 channel analog-to-digital converter (ADC), and a multi-channel analyzer (MCA) system. The MCA is a commercial microprocessor-based system that is capable of data acquisition and limited data analysis. Its specific data analysis functions include automatic peak location, peak energy, peak area and peak resolution calculations. It can read out to either a teletype or a laboratory minicomputer system.

Peak location and peak area calculations were also performed with a minicomputer system, with a more sophisticated routine than that used with the multi-channel analyzer (6). Data plots were prepared with the minicomputer and associated peripheral equipment. In general, peak area calculations obtained with the multi-channel analyzer system and the minicomputer were in good agreement. For overlapping or close peaks, the minicomputer system was able to calculate peak areas when the multi-channel analyzer-microprocessor calculation failed.

Plutonium Analysis with the Well-Type Ge(Li) Detectors

Both x-rays and gamma-rays are associated with plutonium decay. References 7, 8, and 9 discuss the use of L x-rays for

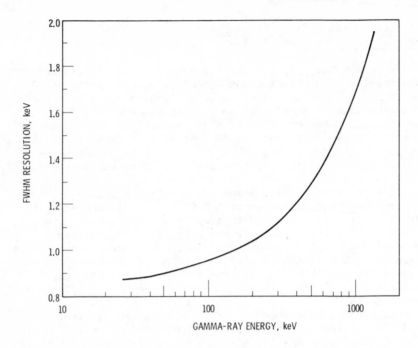

Figure 1. Well-type Ge(Li) detector resolution as a function of energy for a point source in the well

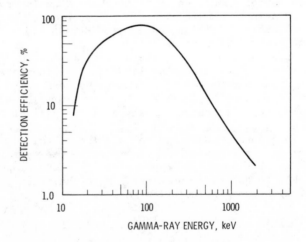

Figure 2. Efficiency (Counts per gamma ray) for a well-type Ge(Li) detector

plutonium assay. The most abundant x-ray associated with the decay of ^{239}Pu is the $UL_{\beta 1}$ x-ray at 17.22 keV. Other less abundant L x-rays range in energy from 11.6 keV to 20.8 keV. UL x-rays are also emitted in the decay of ^{238}Pu and ^{240}Pu. The major interference in L x-ray measurement is caused by the NpL x-rays associated with the decay of ^{241}Am; the energy of the prominent $NpL_{\beta 1}$ x-ray for this isotope is 17.75 keV. Spectra of 99% ^{239}Pu and 99.9% ^{241}Am were accumulated with the source in the well-type Ge(Li) detector (Figure 3). The counting efficiency of the well-type Ge(Li) detector was determined with the high purity ^{239}Pu and ^{241}Am sources. Table I summarizes this data and associated backgrounds for the most prominent spectral peaks and also indicates the abundances of ^{238}Pu and ^{240}Pu $UL_{\beta 1}$ x-rays and gamma-rays.

The resolution of the Ge(Li) well detector is not adequate to separate the $UL_{\beta 1}$ and $NpL_{\beta 1}$ x-rays, as can be done with higher resolutions Si(Li) detectors (7,8). Thus, it is necessary to correct the photopeak at 17.22 keV for samples that contain ^{241}Am. This can be done from either the 59.54 keV or 26.34 keV ^{241}Am gamma-rays. Details for performing these corrections have been reported (9).

A maximum sample volume with the well-type Ge(Li) detector is about one ml. Samples of this size, or smaller, can be analyzed for plutonium without the purification often required for alpha energy analysis or alpha counting. The L_β x-rays have

TABLE I

Well-Type Ge(Li) Detector Counting Efficiency for
Plutonium and Americium Low Energy X-rays and Gamma-rays

ENERGY (keV)	NUCLIDE SOURCE	X-RAY LINE	INTENSITY * PHOTONS / 100 DECAYS	RESOLUTION FWHM (keV)	EFFICIENCY COUNTS / 100 DECAYS	BACKGROUND CPS / keV
17.22	^{238}Pu	$UL_{\beta 1}$	6.00 (6.23)			
17.22	^{239}Pu	$UL_{\beta 1}$	1.90 (2.01)	1.0	0.40	0.0067
17.22	^{240}Pu	$UL_{\beta 1}$	4.97 (5.15)			
17.75	^{241}Am	$NpL_{\beta 1}$	9.56 (12.55)	1.1	3.7	0.0067
26.34	^{241}Am		2.45	0.87	1.14	
38.66	^{239}Pu		0.11	0.90	0.0073	
43.48	^{238}Pu		0.039			
45.23	^{240}Pu		0.045			
51.63	^{239}Pu		0.027	0.90	0.020	0.0031
59.54	^{241}Am		35.9	0.91	27	0.0031

* INTENSITIES FROM REFERENCE (8)

NUMBERS IN ( ) ARE SUMS OF $L_{\beta 5}$ $L_{\beta 1}$ $L_{\beta 3}$

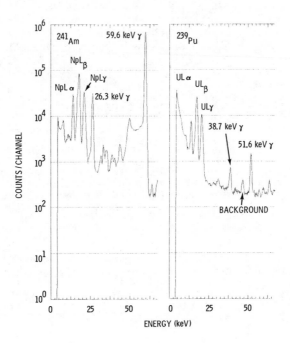

Figure 3. The ^{239}Pu and ^{241}Am spectra taken with a well-type Ge(Li) detector

been detected with the well-type Ge(Li) system in samples containing as little as 2 disintegrations per second of ^{239}Pu.
Additional research is needed to determine the precision and
accuracy of such low-level measurements and the effects of various levels of ^{241}Am and solution densities.

The application of gamma-ray spectrometry to plutonium isotopic analysis has been reviewed by several authors (10,11,12,13,
14). The well-type Ge(Li) detector has resolution similar to,
and efficiency higher than, the planar or coaxial detectors
normally used for plutonium isotopic analysis. Thus, smaller
quantities of plutonium are required for isotopic analysis with
this type of detector than with Ge(Li) detectors of other geome-
tries. Several samples of plutonium with different isotopic
ratios and varying amounts of ^{241}Am were counted in the well-type
Ge(Li) detector. The spectra obtained indicate that with adequate
calibration, satisfactory plutonium isotopic analyses can be
performed with this detector. A spectrum of a plutonium sample
(NBS SRM-948) is shown in Figure 4. Resolution of peaks used for
isotopic analysis is also shown.

Uranium Analysis with Ge(Li) Detector

The use of the well-type Ge(Li) system for uranium analysis
was evaluated with three different types of materials: National
Bureau of Standards (NBS) isotopic standards (U_3O_8) of approximately 0.1 g size, a series of 2 µg uranium solution samples
prepared from the NBS isotopic standards, and a series of New
Brunswick Laboratory (NBL) uranium ore standards. These samples
were selected to represent the application of well-type Ge(Li)
gamma-ray spectrometry to uranium feed and product material
analysis, uranium waste loss analysis, and ore sample analysis,
respectively.

A portion of the spectrum of a 1% ^{235}U NBS U_3O_8 standard is
shown in Figure 5. The gamma-rays associated with the decay of
uranium that are most useful for gamma-ray spectrometric measurements are the 163.4 keV and 185.7 keV gamma-rays from ^{235}U, the
63.3 keV gamma-rays from the ^{234}Th daughter of ^{238}U, and the
1001 keV ^{234}Pa gamma-ray (not shown in Figure 5) from the decay
of ^{238}U.

From the gamma-ray spectrometric measurements made on the
NBS U_3O_8 isotopic standards, observed photopeak areas in counts
per second (CPS) per gram of ^{238}U and mg of ^{235}U are listed in
Table II. Count rates per gram of isotope for a given photopeak
were found to be constant regardless of ^{235}U enrichment. Some
variability associated with the differences in total sample
weight was observed; use of a constant sample size is necessary
to reduce self-absorption effects, especially at low energies.
The ^{238}U detection efficiency was found to be highest with the
63.3 keV ^{234}Th photopeak. Self-absorption effects were less with
the higher energy 1.001 MeV ^{234}Pa photopeak. For ^{235}U, the

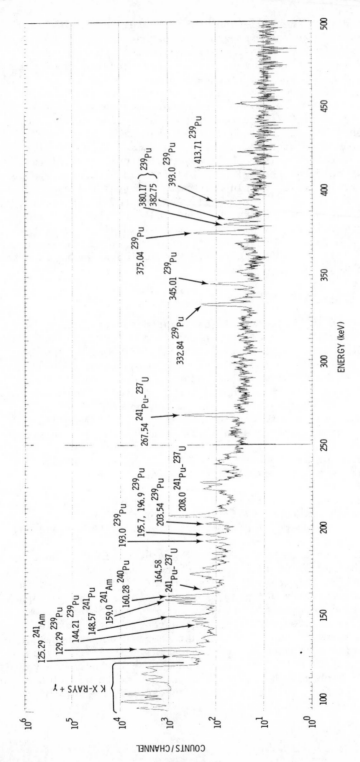

Figure 4. Spectrum of SRM 948 Pu isotope standard in a well-type Ge(Li) detector

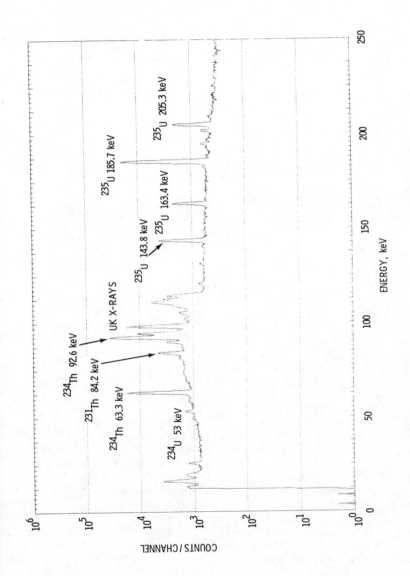

Figure 5. Spectrum of U_3O_8 (1% ^{235}U) in a well-type Ge(Li) detector

TABLE II

Well-Type Ge(Li) System Measurements
of NBS U_3O_8 Isotopic Standards

SAMPLE		CPS/^{238}U, g		CPS/^{235}U, mg	
U, g	^{235}U, at.%	^{234}Th, 63.3 keV	^{234}Pa, 1000.1 keV	^{235}U, 163.4 keV	^{235}U, 185.7 keV
0.123	1.00	91	6.3	1.10	12.3
0.0758	2.04	109	6.3	1.25	13.6
0.145	5.01	80	6.3	1.02	11.8
0.117	15.31	88	6.2	1.09	12.1
0.0947	35.19	90	5.8	1.03	11.6
0.0851	75.36	98	5.9	1.14	12.6
0.0878	85.14	106	5.6	1.19	13.0
0.110	93.34	110	6.5	1.01	11.2
BACKGROUND (CPS)		0.0053	0.00026	0.0026	0.0054

highest detection efficiency was obtained with the 185.7 keV
photopeak. Aged material containing substantial uranium daughter
activity emits 186.1 keV ^{226}Ra gamma-rays. For such material the
163.4 keV ^{235}U is preferred, since there are no interfering
gamma-rays. Traces of fission product activity interfere with
the 143.8 keV ^{235}U peak.

Results of measurements on solutions containing 2 µg of
uranium in 100 µl of solution are shown in Figure 6. The ^{235}U
content of the solutions varied from 10 to 90%. The gamma-ray
spectrometric method measures the ^{235}U rather than the total
uranium, as is measured by fluorometric methods. This is an
advantage for ^{235}U accountability, especially in enriched uranium
processing facilities where isotopic degradation may occur in the
waste materials. The background on the detector at 185.7 keV
for these measurements was 4.1 counts per 10^3 seconds and could
be further reduced with improved shielding. The precision of the
^{235}U measurements is primarily dependent upon the counting time.
This can be adjusted to meet required accuracy. Measurements by
gamma-ray spectrometry are not subject to many of the interference uncertainities associated with chemical methods for low-
level uranium analysis.

Results of analyses on a series of ore standards obtained
from NBL are plotted in Figure 7 for five different gamma-ray
photopeaks. Uranium contents of the samples ranged from 0.025%
to 45%. Only the monazite sand sample containing 9% thorium gave

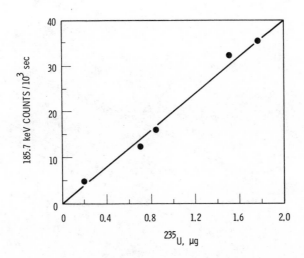

Figure 6. Well-type Ge(Li) detector measurements of dilute uranium solutions of various ^{235}U contents (total $U = 2\mu g/$ sample; $^{235}U = 10–90\%$)

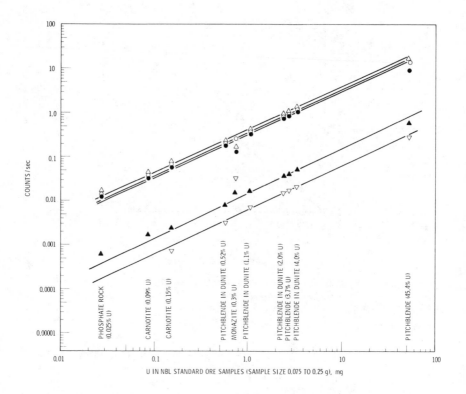

Figure 7. Well-type Ge(Li) detector measurements of NBL ore standards with various U concentrations (●, 63.3 keV; △, 92.6 keV; ▲, 163.4 keV; ○, 185.7 keV; ▽, 1001.0 keV)

spurious results at most of the photopeaks, and the most intense
185.7 keV gamma-ray gave reliable results for this sample. In
case there is a disequilibrium between the daughters and the
parents the use of several photopeaks for analysis of ore samples
is desirable. The 163.4 keV ^{235}U peak would generally not be
affected by daughter activities.

Discussion

A potential application of Ge(Li) well-type detectors in
nuclear safeguards accountability is the measurement of low-level
plutonium and uranium in waste streams. To evaluate the detector
for this application we calculated minimum detectable activity
levels for different counting times based on the calibration data
reported above and according to the method presented by Walford,
Cooper, and Keyser (15). A value of 0.33 for the peak area
standard deviation divided by the peak area was used in the compu-
tations. Samples were assumed to be free of interfering activities which increase the background under the photopeaks. The
minimum detectable activities for various counting times are
listed in Table III. The plutonium values are based upon pure
^{239}Pu. Slightly different values would be obtained for other
plutonium isotopic compositions. To obtain accurate data,
calibrations must be performed with sources where isotopic compositions are near those of the material being processed at the
facility. Higher sensitivities for plutonium analyses can be

TABLE III

Minimum Detectable Activity for ^{239}Pu and ^{235}U
Measured With a Well-Type Ge(Li) Spectrometry System

	MINIMUM DETECTABLE ACTIVITY		
COUNTING PERIOD (SECONDS)	^{239}Pu UL$_{\beta_1}$ X-RAYS (ng)	^{239}Pu 51.63 keV GAMMA-RAY (ng)	^{235}U 185.7 keV GAMMA-RAY (μg)
10^2	13	240	6
10^3	2.6	40	1.1
10^4	0.7	10	0.3
10^5	0.2	3	0.09

obtained by analyzing ^{241}Am. However, it is then necessary to
assume a constant ratio between the ^{241}Am and the plutonium. The
uranium detection sensitivity is comparable to that obtained by
other analysis methods.

We compared peak area calculations by several methods in
this investigation. These included the calculation capabilities
of a commercial microprocessor-based multichannel pulse height
analyzer, with a built-in peak area computation routine, and more
powerful peak area calculation routines of a laboratory minicomputer. The minicomputer results agreed with the hardwired
(microprocessor) system for the most prominent peaks. However,
for overlapping or relatively low abundance peaks, the minicomputer routine was found necessary for reliable results. A
laboratory system based on a modern microprocessor-type multi-
channel analyzer could be used for routine analyses, but careful
selection of peaks and calculation parameters is necessary.
Direct read-out of concentration is possible with such systems.

Acknowledgments

The authors wish to acknowledge the assistance of S. C.
Simpson, who performed many of the experimental measurements
for this investigation.

[1]This paper is based on work sponsored by the Division of Chemical
Sciences, United States Department of Energy, under Contract
EY-76-C-06-1830.

Literature Cited

1. Rodden, C. J., "Selected Measurement Methods for Plutonium
   and Uranium in the Nuclear Fuel Cycle," U.S. Atomic Energy
   Commission, TID-7029, 1972.
2. Reilly, T. D. and Evans, M. L., "Measurement Reliability for
   Nuclear Material Assay," Los Alamos Scientific Laboratory
   Report, LA-6574, 1977.
3. Hakkila, E. A., et al., "Coordinated Safeguards for Materials
   Management in a Fuel Reprocessing Plant II. Appendix, "Los
   Alamos Scientific Laboratory Report, LA-6881 Vol. II, 1977.
4. Shipley, J. P., et al., "Coordinated Safeguards for Material
   Management in a Mixed-Oxide Fuel Facility," Los Alamos
   Scientific Laboratory Report, LA-6546, 1977.
5. Brauer, F. P. and Mitzlaff, W. A., "Evaluation of Well-Type
   Ge(Li) Detector for Low-Level Radiochemical Analysis," IEEE
   Trans. Nucl. Sci., (1978) NS-25.(1) 398-403.
6. Brauer, F. P. and Fager, J. E., "A Computer System for
   Environmental Sample Analysis and Data Storage and Analysis,"
   IEEE Trans. Nucl. Sci., (1976) NS-23 (1) 743-747.
7. Brauer, F. P., Kelley, J. M., Goles, R. W. and Fager, J. E.,
   "Measurement of Environmental ^{241}Am and the Pu/^{241}Am Ratio by
   Photon Spectrometry," IEEE Trans. Nucl. Sci., (1977) NS-24 (1)
   591-595.

8. Strauss, M. G., Sherman, I. S., Swanson, E. J.,"Considerations in Measuring Trace Radionuclides in Soil Samples by L X-ray Detection," IEEE Trans. Nucl. Sci., (1978) NS-25 (1) 740-756.
9. Nielson, K. K., Thomas, C. W., Wogman, N. A. and Brodzinski, R. L.,"Development of a Plutonium-Americium Monitor for In Situ Soil Surface and Pond Bottom Assay," Nucl. Instr. and Meth., (1976) 138 227-234.
10. Parker, J. L. and Reilly, T. D., "Plutonium Isotopic Determination by Gamma-Ray Spectroscopy," Los Alamos Scientific Laboratory Report, LA-5675-PR, 1974, 11-19.
11. Haas, F. X., and Lemming, J. F., "Gamma-Ray Isotopic Measurements for Assay of Plutonium Fuels," J. Institute of Nucl. Mat'ls. Management, (1976), Vol. V, No. III, 189-198.
12. Gunnink, R., "Status of Plutonium Isotopic Measurements by Gamma-Ray Spectrometry," Lawrence Livermore Laboratory Report UCRL-76418, 1975.
13. Gunnink, R., Evans, J. E., and Prindle, A. L., "A Reevaluation of the Gamma-Ray Energies and Absolute Branching Intensities of ^{237}U, ^{238,239,240,241}Pu, and ^{241}Am," Lawrence Livermore Laboratory Report, UCRL-52139, 1976.
14. Radu, D. and Pozzi, F.,"Plutonium Isotopic Composition Measurement by High Resolution Gamma-Spectrometry," J. Radioanal. Chem. (1977) 36 537-545.
15. Walford, G. V., Cooper, J. A. and Keyser, R. M.,"Evaluation of Standardized Ge(Li) Gamma-Ray Detectors for Low-Level Environmental Measurement," IEEE Trans. on Nucl. Sci. (1976) NS-23 (1) 734-742.

RECEIVED JUNE 26, 1978.

11

A Portable Calorimeter System for Nondestructive Assay of Mixed-Oxide Fuels

C. T. ROCHE and R. B. PERRY
Special Materials Division, Argonne National Laboratory, Argonne, IL 60439

R. N. LEWIS, E. A. JUNG, and J. R. HAUMANN
Electronics Division, Argonne National Laboratory, Argonne, IL 60439

A. Introduction

The use of calorimeters to perform a nondestructive determination of the plutonium content of nuclear materials has been well documented.(1, 2, 3) The technique is both highly sensitive and precise. Calorimeters are capable of detecting plutonium concentrations in the 20-ppm range. This corresponds to 0.1 $\mu W/cm^2$, or approximately 0.1 g $^{239}Pu/liter$. Precision better than 0.1% is obtainable on the measurement of plutonium produced power.(4) Among the advantages of calorimetry, when compared with other nondestructive-analysis techniques, are its insensitivity to the chemical form of the plutonium and its independence of measurement-bias problems due to sample geometric configuration and sample matrix composition. However, the lack of portability of the instrumentation and the relatively long sample assay time, when compared to neutron and gamma-ray procedures, have prevented calorimetry from being used as an in-field analytic technique. In designing ANL air-chamber isothermal calorimeters, we have constructed low-thermal-capacitance devices which eliminate the necessity for the large water-bath heat sinks used by classical heat-flow calorimeters. The first of a set of instruments designed to assay the types of plutonium-containing materials encountered by International Atomic Energy Agency (IAEA) inspection personnel has been completed and tested. This device, the "Small-Sample Calorimeter," has been designed to measure the plutonium content of fuel pellets, powders, and solutions. The reduction in instrument size and mass, coupled with the improvement in measurement time should enable inspectors to employ this technique for in-field verification of nuclear safeguards systems.

B. Radioactive Decay and Sample Specific Power

Calorimetry is a technique which can be used to measure the thermal power produced by decaying radionuclides. This power is related to the nuclide mass through the total decay energy and the specific activity of the disintegration. The nuclear constants for the isotopes present in mixed-oxide fuel which are

of importance to calorimetric assay are listed in Table I.(4)
These include the plutonium isotopes (A = 238-242) and ^{241}Am.
The uranium isotopes present in MOX fuel are not included since
the power produced by these nuclides is insignificant when compared to plutonium.

With the exception of ^{241}Pu, the principal decay mode of
the isotopes in Table I is alpha-decay, with a total disintegration energy between 5-6 MeV. Alpha particles in this energy
interval are very short-ranged (< 50 mg/cm^2 of Al). The other
major radiations produced by these isotopes, a 5.5 keV beta from
^{241}Pu and a 59.5 keV gamma from ^{241}Am, are also short-ranged
and, like the alpha's, will deposit their energy within the sample or the measurement-chamber walls. Sources of energy which
will not be measured by the calorimeter include high energy gamma-
-rays, neutrinos, and neutrons; however, these account for less
than 0.01% of the total decay energy. Consequently, calorimetry
is a technique for which the theoretical detection efficiency
approaches 100%.

In order to convert the calorimetrically measured wattage
into plutonium content, it is necessary to determine independently
the power emitted per gram of sample within a particular isotopic
composition. The power emitted by a sample composed of a single
isotope is:

$$P_i = 1.1167 \times 10^9 \times Q \times \lambda \times M_i/A$$

where

P_i	—	specific power of the isotopic i (mW/g)
Q	—	total disintegration energy (MeV)
λ	—	decay constant (day^{-1})
A	—	gram atomic weight of isotope i (g)
M_i	—	mass of isotope i (g)

Most material encountered by IAEA inspection personnel during
an assay contains a mixture of the isotopes in Table I. Thus,
the sample power would be:

$$P_S = \sum_i^n M_i P_i = M_T \sum_i^n R_i P_i = M_T P_{eff}$$

where

P_S	—	sample power (mW)
M_T	—	total sample mass of Pu (g)
R_i	—	mass fraction of isotope i
P_{eff}	—	effective specific power (mW/g)

Typical isotopic composition of nuclear fuels which may
be encountered are given in Table II.(5) The principal constitu-
ent is ^{239}Pu which varies approximately 30% over the fuels listed.
The variation in the other principal heat-producing nuclides
is larger. Plutonium-240 ranges over a factor of 4, and ^{238}Pu

TABLE I.

NUCLEAR CONSTANTS FOR RADIONUCLIDES IN PU-BEARING MATERIALS [1]

Nuclide	Q Total Disintegration Energy (MeV)	Principle Decay Mode	%Q Emitted as γ	$T_{1/2}$ (Years)	λ (days^{-1})	Specific Power mW/g
^{241}Am	5.6402 ± 0.0016	α	0.38	434.1 ± 0.6	4.372×10^{-6}	114.23 ± 0.16
^{238}Pu	5.5921 ± 0.0016	α	0.0005	87.79 ± 0.08	2.1617×10^{-5}	567.16 ± 0.57
^{239}Pu	5.2428 ± 0.0016	α	0.0003	24082. ± 46	7.880×10^{-8}	1.9293 ± 0.0053
^{240}Pu	5.2551 ± 0.0016	α	0.0005	6537. ± 10	2.903×10^{-7}	7.098 ± 0.015
^{241}Pu	0.00553 ± 0.00001	β	0.0045	14.35 ± 0.02	1.322×10^{-4}	3.39 ± 0.002
^{242}Pu	4.985 ± 0.01	α	0.0	370399. ± 2900 (1σ)[2]	5.01×10^{-9}	0.118 ± 0.001

[1] ANSI N15-22-1975

[2] J. M. Meadows, ANL/MDM-38 Half-life determined relative to $T_{1/2}$ (α) of 24290 ± 70 yrs for ^{239}Pu

TABLE II.

RADIONUCLIDE ABUNDANCE OF TYPICAL FUELS

wt %

Fuel Type

Nuclide	LWR	FBR (ZPPR)		British	Recycle	WR
		F & G	H			
^{241}Am	0.7	0.15	--	0.1	1.5	0.05
^{238}Pu	0.2	0.05	0.09	0.1	0.5	0.01
^{239}Pu	75.7	86.6	68.5	90.2	65.0	93.5
^{240}Pu	18.4	11.5	25.6	8.5	24.0	6.0
^{241}Pu	4.6	1.7	4.53	1.0	8.0	0.5
^{242}Pu	1.1	0.2	1.4	0.2	2.5	0.05
P_{eff} (mW/g)	4.86	2.96	3.80	3.06	7.78	2.36

a factor of 10. Americium-241 is the decay product of ^{241}Pu,
and, consequently, the americium content of the sample is dependent upon the length of time since the fuel was reprocessed.
As a result of the ingrowth of ^{241}Am, the effective specific
power of a fuel sample will increase with time. The effective
specific power in mW/g of Pu is also given in Table II.

The effect of changing isotopic composition on the relative
heat contribution of the isotopes in MOX-LWR (light water reactor) fuel is shown in Figs. 1, 2.(9) As fuel burn-up proceeds,
the relative amounts of the nuclides with large specific powers
increase. Consequently, the P_{eff} for a particular fuel may vary
significantly depending upon its residence time in the reactor.

The effective specific power may be determined in two ways.
The first method, referred to as the empirical technique, requires that a set of representative samples be collected. These
samples are measured calorimetrically, and a subsequent chemical
assay is performed to determine their plutonium content. A time
spread in the analysis of the sample is necessary to determine
the change in specific power due to the ingrowth of ^{241}Am. In
general, this technique is not suitable for inspection personnel
because of the time delay required to account for the americium.
The second method, referred to as the computational technique,
requires that an isotopic analysis be performed and that the mass
ratios be used to calculate a weighted-average specific power.
The isotopic analysis may be performed either by mass-spectrometric or NDA techniques. The total assay may be performed
nondestructively by combining a gamma-spectrometric analysis
of the sample isotopic with a calorimetric measurement. A 100-
min gamma analysis which uses a version of GAMANL, a gamma-
stripping program designed for analysis of plutonium,(7) gives
the specific power with a precision of better than 1% for a sample containing 1-2 g of plutonium.

After determining the P_{eff} for the representative material,
the mass of other samples may be determined from the relation

$$\bar{M}_S = \bar{P}_S/P_{eff}$$

The uncertainty in the sample mass is given by

$$(\sigma(\bar{M}_S)/\bar{M})^2 = [\sigma(\bar{P}_S)/\bar{P}_S)^2 + (\sigma(P_{eff})/P_{eff})^2]$$

where $\sigma(\cdot)$ is the standard deviation of the measurement.

C.   Air-Chamber Calorimeters
The classical heat-flow calorimeter measures the temperature
across a thermal resistance for a plutonium-containing chamber
in contact with a large isothermal heat sink. The entire system
is allowed to reach a steady-state condition in which the heat
absorbed by the water-bath heat sink is equal to the heat produced by the decaying radionuclides. The plutonium-produced
power is then proportional to the difference between the sample
chamber temperature and the heat-sink temperature. The ANL air-

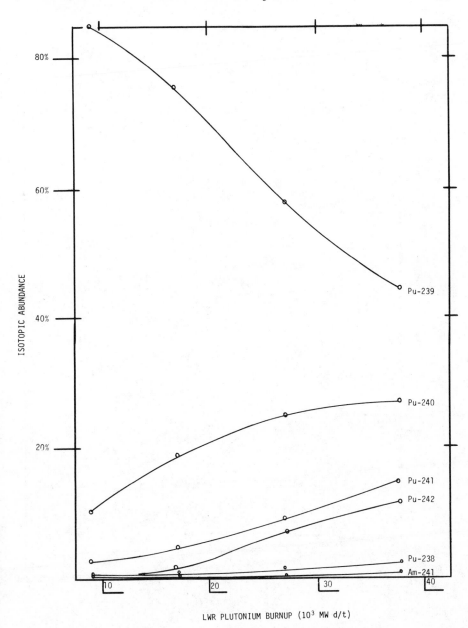

Figure 1. Isotopic abundances of a mixed-oxide, light-water reactor (MOX–LWR) fuel as a function of irradiation history

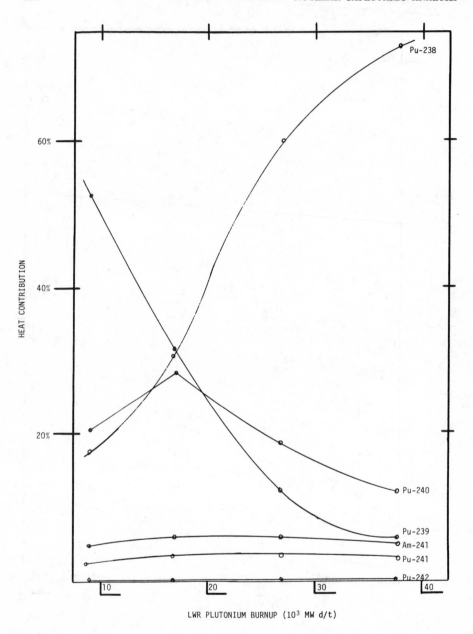

Figure 2. *Heat contribution of the isotopes in MOX–LWR fuel as a function of radiation history*

chamber calorimeters are constructed from a series of concentric cylinders which act as a constant-temperature oven. The servo circuits measure the electrical power necessary to maintain this constant temperature. This design eliminates the necessity for a large isothermal heat sink. The temperature-power relations in the air-chamber calorimeter are shown in Fig. 3. At equilibrium an amount of calorimeter-supplied electrical power, P_0, is necessary to maintain the measurement chamber at some constant temperature, T_3. If a fuel sample at some temperature less than T_3 is inserted into the measurement chamber, the heaters will apply power to raise the combined sample-sample chamber system to T_3. The total power necessary to reestablish T_3 is P_0; however, the amount of power supplied by the control circuits, P_C, depends upon the sample composition. If the sample is producing heat due to radioactive decay, then P_C will be less than P_0, and the sample power may be determined by the relation

$$P_S = P_0 - P_C$$

Table III shows the accuracy attainable with this technique. In these experiments, ZPPR fuel rods were assayed by neutron-coincidence, gamma-assay, and calorimetric techniques.(6,8) The results were compared with a chemical analysis of representative fuel rods. The calorimeter used in these experiments was the ANL Model II, which had an assay time of 20 minutes. The calorimetric results agree with the chemical analysis within experimental uncertainty, while the neutron and gamma assays seem to display a measurement bias. This is a situation where the accuracy of the more rapid assay techniques can be improved by a concurrent calorimetric assay.

D. <u>Small-Sample Calorimeter</u>
The small-sample calorimeter is shown in Fig. 4. The system consists of two instrument packages: a measurement module and a data-acquisition module. The combined weight of both packages is 18 kg. The small-sample calorimeter is capable of measuring samples producing thermal power up to 32 mW. This is equivalent to approximately 10 g (6 cc) of plutonium oxide. This device has a measurement cycle of 20 min with a precision of 0.1%.
The data-acquisition system (DAS) is housed in a 47 cm × 35 cm × 16 cm attaché case and has a weight of 5 kg. It is a totally dedicated microprocessor-controlled device designed around the Intel 8085. The DAS obtains calorimetric power data through a 13-bit ADC. The system memory consists of 8-K bytes of erasable programmable read-only memory (EPROM) and 1-K bytes of random-access memory (RAM). The system program resides in the permanent memory (EPROM) and does not need to be reentered after the device is powered down. The upper 2-K bytes of this nonvolatile memory are located on an EPROM chip residing in the zero-insertion-force socket on the face of the unit. The EPROM

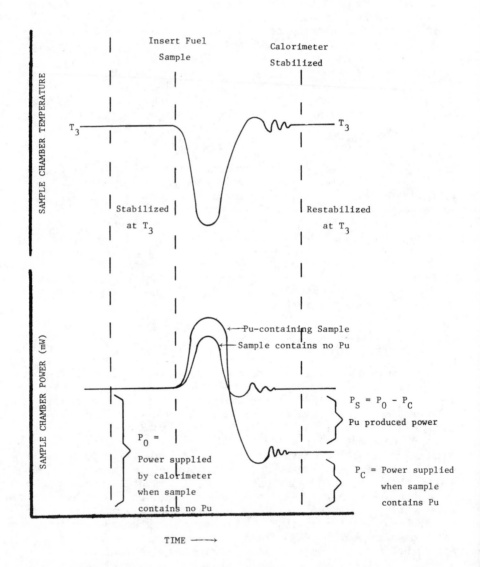

Figure 3. Measurement-chamber temperature and power relations in ANL air-chamber calorimeters

TABLE III,

COMPARISON OF ANL-TYPE CALORIMETER WITH
OTHER ANALYTICAL TECHNIQUES (8)

Physical composition: 0.95 cm × 15.25 cm stainless steel clad fuel rods
Chemical composition: PuO_2, UO_2
Nominal isotopic composition:

Rod type	^{238}Pu	^{239}Pu	^{240}Pu	^{241}Pu	^{242}Pu
F, G	0.05	86.55	11.5	1.7	0.2
H	0.09	68.45	25.56	4.53	1.38

Plutonium wt %

Technique	F	G	H
Neutron coincidence	13.35 ± 0.06	26.26 ± 0.07	NA
Gamma-ray assay	13.28 ± 0.03	26.42 ± 0.04	NA
Model II Calorimeter	13.20 ± 0.03	26.51 ± 0.03	15.73 ± 0.01
Chemistry	13.19 ± 0.02	26.48 ± 0.04	15.76 ± 0.06

Figure 4. ANL small-sample calorimetric system

chip residing in this socket may be exchanged for other chips
containing different programs. This permits the user to develop
data-handling and statistical programs tailored to his specific
needs.

The software provided with the unit includes a number of
data-acquisition and -handling routines. These include codes
which calculate the average measurement-chamber electrical power
($\bar{P}$) and its standard deviation. The DAS will store the empty
chamber baseline power for comparison with results obtained
during a sample assay. Analysis programs are also included to
obtain the sample mass and its uncertainty from the power measurement and the effective specific power. There is also a routine to calculate the P_{eff} from the sample isotopic data, as
well as a routine to correct mass-fraction data for changes due
to radioactive decay.

A double-encapsulation technique using metal sample containers was chosen to maximize the rate of heat transfer (see
Fig. 4). The inner capsule is an inexpensive, commercially
available drawn-aluminum cylinder with a diameter of 1.6 cm and
a length of 5 cm. It would be sealed inside a glove box after
the MOX powder or pellets were inserted and could be disposed
of after the assay. The outer sample holder has been machined
to minimize the air gap between the inner capsule and the sample-
chamber walls. This cylinder has outer dimensions of 2 cm in
diameter and 7 cm in length. It is fitted with an "O" ring seal
to minimize the chance of radiation contamination. By the use
of this encapsulation technique, a measurement precision of 0.1%
can be obtained in an equilibration time of 15 min. Equilibration tests were also conducted on sources double-bagged in polyethylene. The poor heat-conducting properties of the plastic
slowed the equilibration time to about 30 min.

The measurement module contains the calorimeter chamber,
the sample preheater, and the control circuits. Figure 5 shows
the relation between the cylinders comprising the calorimetric
unit and the measurement and control circuitry. The small-sample
calorimeter is constructed of four concentric shells which are
maintained at progressively higher temperatures approaching the
center of the unit ($T_{ROOM} < T_0 < T_1 < T_2 < T_3$). This ensures
that heat flow will remain constant in the outward direction
across the temperature sensing coils. The outer shells, T_0 and
T_1, act as protective buffers for the inner measurement cylinders. They are controlled by a servo circuit which has YSI ther-
mistors as sensors and copper coils wound around the aluminum
cylinders as heaters. The temperature difference between the
inner measurement cylinders, T_2 and T_3, must be controlled to
the microdegree range. The Ni coils on these cylinders act as
both temperature sensors and heaters by using the principles
of resistance thermometry.

A simplified diagram presenting the control principles used
in these circuits is shown in Fig. 6. The control circuits con-

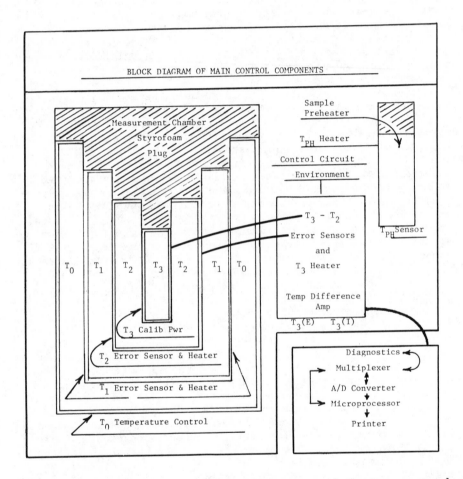

Figure 5. Block diagram of the small-sample calorimeter with its measurement and control components

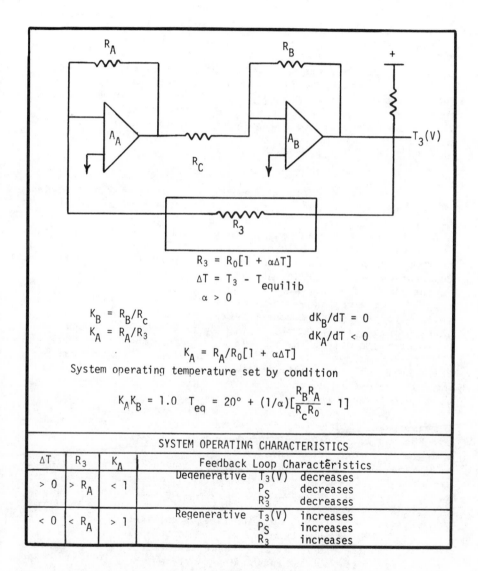

Figure 6. Resistance thermometry and feedback-control circuitry in ANL air-chamber calorimeters

sist of two operational-amplifier (OP-AMP) networks. AMP·A
senses the resistance imbalance between a precision resistor
(1 ppm/°C) R_A and the Ni coil, R_3, wound around the measurement
chamber. The resistance of R_3 is dependent upon the temperature

$$R_3 = R_0[1 + \alpha T]$$

where
   R_0 –   resistance at 20°C
   α –   temperature coefficient of resistance (> 0)

AMP·B acts as a booster which determines the amount of power
that will be applied to R_3. This circuit forms a negative feedback system. If T_3 is too low, R_3 will be less than R_A. Consequently, the gain of AMP·A will be greater than 1.0. This
will cause more power to be applied to R_3 which will raise the
temperature in the measurement chamber. The resistance of R_3
will increase until it equals R_A.

E.   Data Analysis
     The procedures followed during a typical assay are included
in the flowchart shown in Fig. 7. One of the more important
features of the calorimetric technique is that the operator may
calibrate the instrument using electrical heat standards. This
is especially vital for in-field use of the device, since severe
restrictions have been placed on the transportation of plutonium
calibration sources. The DAS will automatically perform the
calibration for a user-selected number of input calibration
powers. The microprocessor calculates the proper input reference
voltage to be applied across the calibration resistance coil.
These powers simulate a set of plutonium standards over the measurement range of the instrument. The system then measures both
the input power and the control-circuit power at each point.
The results of this calibration are used in a linear least squares
analysis to determine the zero-power intercept (A) and the power-
measurement slope (B).
     An assay is performed with a Pu standard to determine if
a measurement bias exists between the electrically produced power
and the radioactive-decay produced power. The results are incorporated in the normalization constant F.
     If a number of samples are to be assayed, the operator will
periodically check the system stability by measuring the control-
circuit supplied power for samples which do not contain a heat-
producing source (P_0). A t-test comparison may then be performed
to assure that there has been no significant electronic drift
from the calibration-determined empty-chamber power (A).
     The power produced by an unknown source is then determined
by the use of the above-mentioned parameters

DATA ANALYSIS FLOWCHART

Electrical Calibration

1. Apply reference voltage (P_A)
2. Monitor calorimeter output (P_M)
3. Calculate A, B (LLSQ)

$$\bar{P}_M = A + B\bar{P}_A$$

Plutonium Calibration

1. Assay calibrated standard (P_K)
2. Calculate $F = P_{STND}/P_K$

System Stability

1. Measure P_C for sample containing no Pu (P_0)
2. t-test comparison of P_0 with A

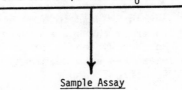

Sample Assay

1. Measure P_C for unknown source
2. Calculate $\bar{P}_S = F(\bar{P}_C - A)/B$
3. Calculate *
   $s^2(\bar{P}_S) = s^2(HDE) + s^2(TEMP) + s^2(REP) + s^2(CALIB) + s^2(\bar{P}_C)$
4. Calculate $M_S = \bar{P}_S/P_{eff}$
5. Calculate
   $s^2(M) = s^2(\bar{P}_S) + s^2(P_{eff})$

*$s(\cdot)$ represents the relative standard deviation of tne measured quantity.

Figure 7. Data analysis flowchart for the small-sample calorimeter

$$\overline{P}_S = F(\overline{P}_C - A)/B$$

The uncertainty associated with this power measurement includes
the following sources of error:
1. $s(HDE)$ – the heat-distribution error contains the uncer-
    tainty in the measurement precision due to effects of geo-
    metric position of the heat source in the sample chamber.
2. $s(Temp)$ – the ambient-temperature error contains the uncer-
    tainty due to fluctuations in the room temperature during
    the assay.
3. $s(Rep)$ – the sampling-reproducibility error is determined
    by repeated measurements of the power from a single source.
4. $s(Calib)$ – the calibration error includes the uncertainties
    in the electrical and radioactive calibrations.
5. $s(\overline{P}_C)$ – the statistical uncertainty is defined as the stan-
    dard deviation of the power measurement $\div \sqrt{n}$.
$(s(\cdot)$ refers to the relative standard deviation of the quantity
measured.)
The unknown plutonium content is then obtained by

$$M_S = \overline{P}_S/P_{eff}$$

and the power-measurement uncertainty is combined with the error
in P_{eff} to give the uncertainty in M_S.

F. Results
    The results of an experiment in which a set of plutonium-
containing samples were assayed by the small-sample calorimeter
are given in Table IV. These samples were constructed by placing
encapsulated metal sources in sand to simulate the heat-conduction
properties of PuO_2 powders. The sources were made from a PuAl
alloy with a 98.79%–Pu composition. The isotopic composition
of the sources was determined by a 30-min gamma assay which used
a version of GAMANL specifically adopted for Pu analysis.[7]
P_{eff} was calculated by the computational method discussed in
Section B. The sample power was determined in a 4-min measurement following a 15-min equilibration period. The sample Pu
mass agrees well with the reported book value over the range
of sample powers tested (~ 50% of full scale). The coefficient
of variation for the mass determination was better than 1% in
almost all cases, with the major part of this error resulting
from the determination of P_{eff}. Various device parameters, as
well as the magnitude of the error contributions, are discussed
in Fig. 8.

TABLE IV.

CALORIMETRIC ASSAY OF ZPR-3 PLUTONIUM

Physical form   -   1.6 cc stainless steel encapsulated sources
Chemical composition   -   98.79% Pu, 1.17% Al
Typical isotopic composition   - ^{238}Pu = 0.01%, ^{239}Pu = 95.2%, ^{240}Pu = 4.5%
                                      ^{241}Pu = 0.2%, ^{242}Pu < 0.2%, ^{241}Am = 0.2%

Sample #	P_{eff} (mW/G)[1]	Sample Power (mW)[2]	Sample Mass (g)[3]	Reported Book Value
1	2.48 ± 0.01	4.261 ± 0.005	1.72 ± 0.02	1.72
2	2.49 ± 0.01	3.733 ± 0.004	1.50 ± 0.01	1.50
3	2.48 ± 0.01	4.409 ± 0.005	1.78 ± 0.01	1.76
4	2.51 ± 0.02	4.263 ± 0.005	1.70 ± 0.02	1.71
5	2.48 ± 0.02	8.124 ± 0.008	3.28 ± 0.02	3.26
6	2.49 ± 0.02	8.519 ± 0.009	3.42 ± 0.03	3.43
7	2.48 ± 0.02	8.021 ± 0.008	3.23 ± 0.02	3.22
8	2.50 ± 0.02	8.687 ± 0.009	3.47 ± 0.03	3.47
9	2.48 ± 0.02	8.670 ± 0.009	3.50 ± 0.03	3.48
10	2.50 ± 0.02	8.020 ± 0.008	3.21 ± 0.02	3.21
11	2.49 ± 0.03	12.412 ± 0.012	4.98 ± 0.03	4.97
12	2.49 ± 0.03	16.634 ± 0.017	6.68 ± 0.03	6.69

[1] Determined by gamma-assay
    uncertainty contributions - counting statistics, ^{242}Pu bias (< 0.2%)
[2] Determined by 20 min calorimetric analysis
    uncertainty contributions - counting statistics, sample heat distrib. (0.04%)
        system temperature stability (0.09%/°C)
        system reproducibility (0.02%)
[3] Includes 0.1% uncertainty in radioactive standard calibration

ANL SMALL-SAMPLE CALORIMETER

I. Physical Description

    A. Data acquisition module - consisting of 8085 Microprocessor,
        8K-byte nonvolatile memory, printer, and keyboard
           weight - 5 kg
           size   - 47 cm × 35 cm × 16 cm

    B. Measurement module - consisting of calorimeter, sample preheater,
        and power supplies
           weight - 13 kg
           size   - 30 cm × 41 cm × 26 cm

II. Sample Size - up to 10 g (6 cc) of plutonium oxide

III. Environmental Parameters

    A. Line power: 110 VAC (60 Hz) or 220 VAC (50 Hz)
        Sensitivity to line noise - stable with $\pm$ 8 V spikes
        Sensitivity to voltage fluctuations - stable at 110 V $\pm$ 20%

    B. Room temperature operating range - 10.5°C-35.0°C
        Temperature drift < 0.09%/C°

IV. System Equilibration Time (Pre-heated Sample)

    A. Metal encapsulation - 15 min
    B. Double polyethelyne bag encapsulation - 30 min

V.   System Power Measurement Precision

    A. Sample heat distribution uncertainty   -   0.04%
    B. Sampling reproducibility uncertainty   -   0.02%
    C. System temperature stability   -   0.09%/C°
    D. Typical statistical uncertainty
        (4 min assay)   -   0.01%

    Combined precision: $\sigma (P_0)/P_0 = 0.1\%$

Figure 8. Data sheet for the ANL small-sample calorimeter

ABSTRACT

Calorimetric assay provides a precise, nondestructive method to determine sample Pu content based on the heat emitted by decaying radionuclides. This measurement, in combination with a gamma-spectrometer analysis of sample isotopic content, yields the total sample Pu mass. The technique is applicable to sealed containers and is essentially independent of sample matrix configuration and elemental composition. Conventional calorimeter designs employ large water-bath heat sinks and lack the portability needed by inspection personnel. The ANL air-chamber isothermal calorimeters are low-thermal-capacitance devices which eliminate the need for large constant-temperature heat sinks. These instruments are designed to use a feedback system that applies power to maintain the sample chamber at a constant electrical resistance and, therefore, at a constant temperature. The applied-power difference between a Pu-containing sample and a blank determines the radioactive-decay power. The operating characteristics of a calorimeter designed for assaying mixed-oxide powders, fuel pellets, and Pu-containing solutions are discussed. This device consists of the calorimeter, sample preheater, and a microprocessor-controlled data-acquisition system. The small-sample device weighs 18 kg and has a measurement cycle of 20 min, with a precision of 0.1% at 10 mW. A 100-min gamma-ray measurement gives the specific power with a precision of better than 1% for samples containing 1-2 g of plutonium.

178   NUCLEAR SAFEGUARDS ANALYSIS

LITERATURE CITED

1. Gunn, S. R., Nuclear Instruments and Methods (1967) 29, 1.
2. Beyer, N. S., Lewis, R. N., and Perry, R. B., Nuclear Materials Management (1972) 1, (3) 170.
3. Oeting, F. L., J. Inorg. Nucl. Chem. (1965) 27, 151.
4. ANSI N15.22-1975, "American National Standard Calibration Techniques for the Calorimetric Assay of Pu-Bearing Solids Applied to Nuclear Materials Control."
5. Rodenburg, W. W., Proceedings of the Symposium on the Calorimetric Assay of Plutonium (1973) p. 14.
6. Beyer, N. S., Lewis, R. N., and Perry, R. B., Proceedings of the Symposium on the Calorimetric Assay of Plutonium (1973) p. 99.
7. Gunnink, R., "A System for Plutonium Analysis by Gamma-Spectrometry," UCRL-51577 (1974).
8. Beyer, N. S., Perry, R. B., "Four Passive Assay Techniques Applied to Mixed-Oxide Fuel," ANL-7906 (1972).
9. Bishop, D. M. and Taylor, I. N., Proceedings of the Symposium on the Calorimetric Assay of Plutonium (1973) p. 75.
10. Meadows, J. M., "The Alpha and Spontaneous Fission Half-lives of ^{242}Pu," ANL/NDM-38 (1977).

RECEIVED JUNE 5, 1978.

12

Performance of an Accountability Measurement System at an Operating Fuel Reprocessing Facility

M. A. WADE, F. W. SPRAKTES, R. L. HAND, JON M. BALDWIN, E. E. FILBY, and L. C. LEWIS

Allied Chemical Corp., 550 2nd St., Idaho Falls, ID 83401

The Idaho Chemical Processing Plant (ICPP) is a multipurpose facility capable of recovering unfissioned uranium from enriched uranium fuel elements discharged from research, test, propulsion, and power reactors. The ICPP is especially designed to handle fuels with intial enrichments ranging from 20 to 93% ^{235}U. The ICPP has been processing nuclear fuels since 1953 and is currently operated for the Department of Energy by Allied Chemical Corporation.

Throughout its operating history, one of the major concerns at the ICPP has been the maintenance of an accountability measurements system sufficient to generate an accurate and defendable material balance on the process. Within the uncertainties assigned to sampling and to analytical measurements the input and output totals and the process holdup must account for all the material processed, if the material balance is to be regarded as satisfactory.

In 25 years of operation, we have accumulated considerable experience in making and evaluating uranium accountability measurements in an operating plant environment. It is our intention to relate in this paper some of that experience, and to give a comprehensive overview of the uranium accountability measurements system as it currently operates at the ICPP. We will

begin by describing the multiple-fuel process at the ICPP and by
showing the points at which the uranium accountability measurement system interfaces with the process. We will then describe
the components of the measurement system that generate the
accountability information. In that description we will cover
sampling, analytical methodology, calibration, traceability, and
quality control. Finally, we will attempt an assessment of the
overall system performance and will mention areas where improvement is anticipated.

OVERVIEW OF FUEL PROCESSING AT THE ICPP (1)

The ICPP design, in view of its multipurpose aspect, is considerably more complicated than would be encountered in a commercial reprocessing facility. A flowsheet of the ICPP is shown
in Figure 1. The plant features multiple "headends", each
designed for the dissolution of a particular type of fuel. Fuel
types that are currently procesed at the ICPP include enriched
uranium alloyed and/or clad with aluminum, zirconium, or stainless steel. Soon to be operated is a combustion headend for
graphite based fuels.

The usual procedure for aluminum and stainless steel fuels
is to operate in a continuous mode, although batch dissolution
of those fuels can be accomplished. The dissolvent for aluminum
fuels is nitric acid with a Hg(II) catalyst. Stainless steel
fuels are dissolved electrolytically. Zirconium fuels are dissolved semicontinuously by exposing large batches of fuel to a
continuously flowing stream of aqueous HF. In all cases the
entire fuel element, comprised of fuel and cladding, is dissolved. In addition to these more or less standard processes
limited quantities of unique fuels are dissolved in a versatile
custom processing facility.

Aluminum nitrate is added to the zirconium dissolver product
to complex fluoride and reduce corrosion. Whenever possible,
the aluminum dissolver product is used as the source of aluminum. This mode of operation is called coprocessing. Coprocessing gives lower chemical costs, smaller waste volumes, and
higher throughput than can be obtained if cold aluminum nitrate
is used to complex the fluoride.

All dissolver product solutions are decontaminated with a
single solvent extraction cycle using tributyl phosphate in a
kerosene solvent, followed by two cycles of extraction with
methyl isobutyl ketone. Waste streams are monitored for uranium
content and may be recycled, if necessary, to provide adequate
uranium recovery. The product of the extraction process is an
aqueous solution of uranyl nitrate. This aqueous product is
converted for shipment to UO_3 in a fluidized-bed denitration
process.

Incidental to the accountability problem, all fission product wastes are converted by a fluidized bed process to a granular solid oxide matrix and are stored in stainless steel bins
awaiting permanent disposition.

INTERFACE TO THE ACCOUNTABILITY MEASUREMENT SYSTEM

In spite of the complexity evident from Figure 1, the uranium recovery process can be illustrated quite simply for the purpose of defining interface points with the accountability measurement system. This is done in Figure 2, where each operation of fuel reprocessing is shown, irrespective of the particular process vessels and flow paths indicated in Figure 1. The accountability measurement points are indicated by * and are shown to be four in number even though the detailed flowsheet of Figure 1 indicates more than one physical possibility for taking some samples. For example, in the electrolytic dissolution for stainless steel fuels, two input accountability tanks, G-105 and G-155, are available and either one or both may be used in a particular processing run.

Fuel elements are inventoried as discrete items until they reach the dissolver. The dissolver solution is transferred to a calibrated input accountability tank which has facilities for mixing and sampling. Samples from this point are used to establish the input term of the material balance. Waste streams are collected and sampled from tanks similar to the input accountability tanks and measured for uranium concentration and isotopic distribution. The UO_3 final product is sampled as it is loaded into shipment containers and analyzed for uranium concentration and isotopic distribution. Samples from these latter two points form the output term of the material balance. Finally, in order to close a material balance period, the uranium that remains as process holdup must be collected, sampled, and measured. The plant closing inventory for the current material balance period also becomes the starting inventory for the next period.

The characteristics of the samples taken at these points do much to dictate the analytical methodology applied in each particular case. These characteristics are summarized in Table I.

TABLE I. URANIUM ACCOUNTABILITY SAMPLES

Sample Point	Uranium Concentration (g/L)	Activity
Input	0.7 to 20	High
Waste Streams	10^{-3} to 10^{-1}	Low to High
Product	UO_3	Very Low
Inventory	Varies	Moderate to High

All input and inventory samples are corrosive, quite radioactive (up to several R/h/5 mL sample), and moderate to high in uranium concentration. Consequently, these samples must be handled remotely until the fission product activity has been removed, and the uranium content must be measured accurately and precisely. While the product samples are low in activity and

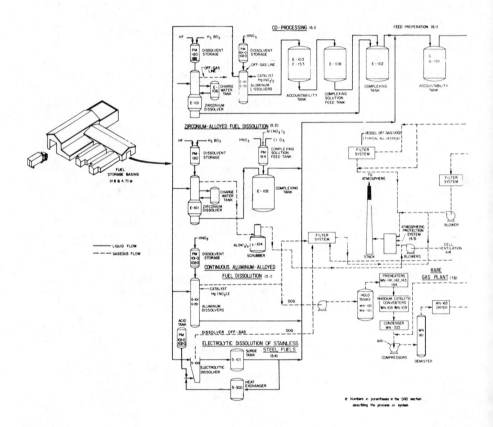

Figure 1. Process flow diagram

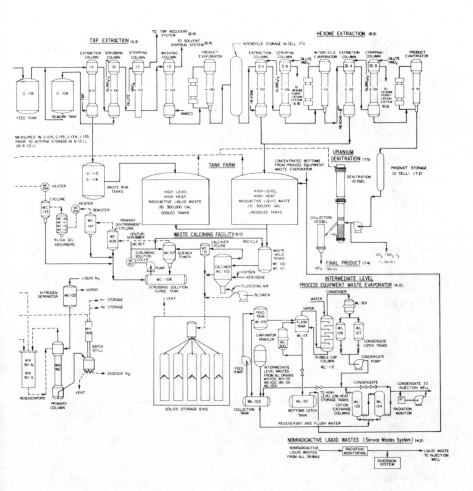

for the ICPP

chemically innocuous the high uranium content again dictates careful measurement. The waste streams are uniquely low in uranium concentration and may generally be analyzed by less precise methods than the other samples. In the next section, we will describe the analytical sequence in detail.

ANALYTICAL SEQUENCE

Input and Inventory Samples

Sampling

The process tanks to be sampled are calibrated by introducing into the tank known mass increments of water and relating these to liquid levels determined from a differential manometer. During sampling for accountability, manometer readings are converted by this established relation to the weight of liquid in the tank. The specific gravity of the process solution is established by laboratory measurement as described in the next section. This measurement provides a check of the homogeneity of the tank contents.

Samples are drawn with air-jet samplers, Figure 3. A 4-mL sample bottle with a septum cap is impaled on two heavy gauge hypodermic needles inside a shielded sample gallery. The sample jet produces a negative differential pressure in the sample return line, causing the tank contents to circulate through the sampling lines and bottle. The tank contents are mixed by air sparging before and during sampling to promote homogeneity. Solution is circulated through the sample bottle for ten minutes to ensure the sample line contents are representative of the tank contents. Three, supposedly identical, samples are drawn for homogeneity verification, and are transported in a shielded container to the analytical laboratory.

Sample Preparation and Homogeneity Checks

As was mentioned earlier, the high level of radioactivity associated with these samples requires that the initial chemical preparation be performed in a shielded enclosure. Thus, preliminary operations on the samples must be designed to require minimum manipulation and to minimize the need for quantitative transfers. This places severe restrictions on the range of analytical techniques that can be used.

Highly radioactive samples are brought into the Remote Analytical Facility (RAF) which consists of 32 boxes, each about one m^3 in size, equipped with manipulators for remote sample handling, and outfitted to perform a specific operation on a sample (Figure 4). The operations pertinent to the treatment of accountability samples are pipetting, weighing, reagent addition, mixing, extraction, and specific gravity measurement.

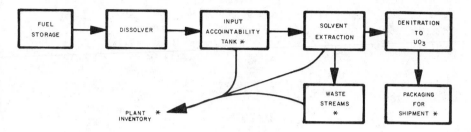

* ACCOUNTABILITY SAMPLING POINTS

Figure 2. Interfaces between the fuel recovery process and the accountability measurements system

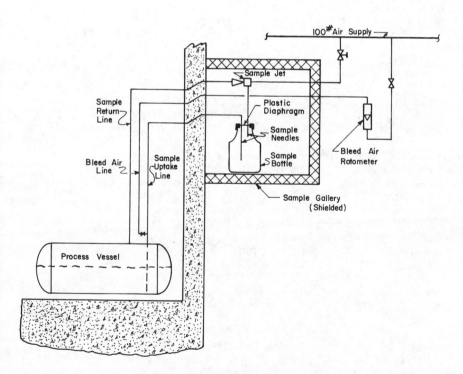

Figure 3. Schematic of a liquid sampler

Specific gravity of the samples is measured with a modified Anton Paar density meter (2). This is a commercial device which measures the shift in resonant frequency of a hollow mechanical oscillator filled with the sample fluid. This shift in resonant frequency is related to the mass of a fixed volume of the fluid by calibration with fluids of known density, usually air and water. The modified unit, shown installed in the shielded cell in Figure 5, is mounted on a temperature controlled metal block and has provision for introducing sample, flushing with a solvent or air, and drying the oscillator tube from outside the cell (3).

The specific gravity of each of the three supposedly identical samples is measured and the range of the three results compared to the standard deviation of the specific gravity measurement as established by the analytical quality control program (q.v.). If the range compares acceptably with the standard deviation of the specific gravity measurement the tank is considered to have been well mixed and the analytical work is continued. If the homogeneity criterion is not met, further mixing and sampling of the contents of the accountability tank is required. The uranium concentration and isotopic measurements are made by an isotope dilution mass spectrometric (IDMS) technique. The measurement is described in a later section. The preparatory steps that must be performed remotely are pipetting, addition of the ^{233}U spike, weighing, and extraction. The high purity spike (99.99% ^{233}U) is prepared by the Quality Control Group and is provided to RAF personnel as individually preweighed aliquots, each containing a known mass of ^{233}U and packaged in a stoppered test tube. A tare weight on this spike tube is taken on a four-place analytical balance that has been modified for remote operation. The spike tube is transferred to a pipettor cell (Figure 6) where an aliquot of the sample (usually 250-400 L) is dispensed into the tube. Design and operation of the remote pipettor has been previously described (4). The tube is then transferred back to the remote balance cell to get a sample weight. The uranium analysis is thus done on a weight basis. The delivered pipettor volume is used only as a check against gross weighing errors by calculating a weight from the delivered volume and the measured specific gravity. Two of the three sample bottles are sub-sampled in this manner for uranium measurement, and the third bottle is held in reserve.

Initial decontamination of the spiked sample is accomplished by a solvent extraction procedure (5,6). Dichromate is added to ensure that the uranium is present only in the +6 oxidation state. A salting solution of "acid deficient" aluminum nitrate and a small amount of aluminon reagent are added to the aqueous phase. The aluminon aids perception of the phase boundary by formation of a red colored complex in the aqueous phase. Observation of the phase boundary through the lead glass window of the cell is otherwise difficult. Methyl isobutyl ketone is

Figure 4. Remote analytical facility

Figure 5. Remote specific gravity instrument

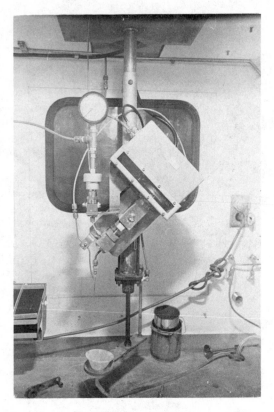

Figure 6. Remote pipettor

added to the sample tube, the tube is stoppered, and placed on a rocker table. After the contents have been adequately shaken, the organic phase is decanted and transferred out of the RAF for further preparative work. The overall decontamination factor for fission products by this method is about 10^5 and the recovery of uranium is on the order of 80-90%.

Solutions formed in the electrolytic dissolution of stainless steel fuels and the dissolution of aluminum fuels require no further decontamination. The uranium is stripped from the organic phase with 0.1 $\underline{M}$ HNO_3, evaporated to dryness, and the residue taken for mass analysis.

Samples from the dissolution of zirconium fuels are decontaminated further by an ion exchange treatment ($\underline{6}$). The organic phase from the extraction is treated with HI to reduce any Pu that might be present, then batch contacted with an aqueous slurry of an anion resin in 8 $\underline{M}$ HCl. The resin is transferred to a short column, washed with 8 $\underline{M}$ HCl, and the uranium eluted with water. The eluate is evaporated to dryness and the residue taken for mass analysis.

Mass Spectrometry

The mass spectrometer used for process support and accountability analyses has been extensively described elsewhere ($\underline{7},\underline{8},\underline{9}$). The instrument is a single-stage $60°$ magnetic-sector mass spectrometer which uses a double-filament thermal ionization source for solid samples and a Faraday cup for ion collection. Samples are inserted into the instrument through a standard Avco vacuum lock and gate valve assembly.

The dried residue from the separation process is first transferred into an alpha-tight glove box. There the solid is redissolved in dilute nitric acid. A few microliters of the solution are loaded onto a tantalum filament and dried by electrical heating. Rhenium is used as the ionizing filament in the center position.

The mass scanning and data collection can be controlled manually or through the computer interface of the system. The dedicated computer system, built around a Raytheon 703 computer, has been described in some detail elsewhere ($\underline{8},\underline{10}$). Initially, the mass range is scanned under manual control with data output to a strip chart recorder. Once a stable beam has been established, scanning and data collection are controlled automatically by the computer. The operator monitors the performance of the system by observing the recorder, the teletype output, and a computer-driven oscilloscope.

The teletype output includes the normalized atom percents for masses-233, -234, -235, -236, and -238. Also listed are the standard deviations of these peaks as determined from the combination of the set of scans taken. The standard deviations of the major peaks are used to establish the acceptability of a set of data. The data set is accepted only if the standard deviations of all the major peaks (defined as 10 atom percent or

greater) are less than some upper limit, currently 0.050.   If
two successive scan sets with major-peak standard deviations
less than 0.05 are not obtained in 45 minutes, the final two
sets are taken as trial data and the sample is reloaded onto a
new filament and rerun.

For accountability samples and related controls, the computer automatically calculates the concentration from the ^{233}U
fraction, the sample weight, and the known weight of spike
added.   The program then renormalizes the intensities for
mass-234 through mass-238 to obtain the atom and weight percent
of uranium in the sample.   In addition to these samples the Mass
Spectrometry Lab runs a daily instrument calibration.   Standards
of known isotopic composition are used and include the National
Bureau of Standard SRM U-500, U-500 mixed with high purity
spike, and others.   The calibration data must agree with known
values and the fractionation correction must agree with the previous geometric average of the fractionation correction within
specified limits.   The geometric average automatically weights
the previous determinations in a decreasing order such that very
old data have essentially no effect on the present value of the
geometric average.

Before this calculation is made, the latest determination
must agree with the previous average within ±0.07 (this limit is
expressed as percent per-mass-unit).   If the proper agreement is
found, the new value will be used throughout that day.   This
calibration must be completed before any accountability samples
or controls can be run.

The acceptability of data is also subject to another con-
straint.   Two out of a set of three samples taken from an
accountability tank are processed through the chemical separation and delivered to the Mass Spectrometry Lab.   The results
for both concentration and the ^{235}U isotopic for these dupli-
cates must agree within limits set by the Quality Control Lab.
If not, the third sample from that tank must be processed and
analyzed.

Waste Stream Measurement

In a complex plant such as the ICPP there are many waste
streams and no attempt will be made to discuss them in detail.
In general, there are two types-those associated with the processing of fuel (raffinates) and those associated with daily
operations (plant waste).

Sampling

In the case of the raffinate streams, two samples are taken
but only one is analyzed initially.   Subsequent action is based
on a comparison of the uranium result with a process limit.   The

result may be reported, the second sample may be analyzed and the two results reported, or additional samples may be requested. The fluorophotometric method (q.v.) which is used for the analyses determines total uranium but the process limit is for ^{235}U. However, the isotopic distribution is available from the analysis of the input sample so the concentration can be corrected to a ^{235}U basis.

The analysis program for the plant waste streams is similar to that for the raffinate streams with the exception that the isotopic distribution of the uranium is not known. The uranium concentration is determined fluorophotometrically. In order to obtain a plant material balance for ^{235}U, samples from the plant waste streams are composited and analyzed on a monthly basis by mass spectrometry.

Fluorophotometric Uranium Measurement

The fluorophotometric method (11,12) is used for the determination of uranium in waste streams because it is highly selective and sensitive to nanogram levels of uranium. It is applicable to a wide variety of aqueous and organic samples containing high levels of both cladding materials and fission products. It is also a fairly simple and a very rapid method; an analysis can be completed in 30 min. The method, as modified for use in our laboratory (6), uses an extraction of the tetrapropyl-ammonium uranyl trinitrate salt into methyl isobutyl ketone from an acid-deficient aluminum nitrate salting solution. The organic phase is evaporated on a sodium fluoride-lithium fluoride pellet which is then fused over a burner. The pellet, after cooling, is irradiated with ultraviolet light at 365 nm from a mercury lamp and the resulting fluorescence, which is proportional to the amount of uranium, is measured at 555 nm. The relative standard deviation of the method is approximately 20%.

Final Product Measurement

Sampling and Archive Samples

As mentioned above, the form of the uranium final product is granular UO_3, prepared from uranyl nitrate solution in a fluidized bed denitrator. UO_3 is transferred by gravity flow from the denitrator to a vee-blender. The product is throughly blended and transferred by gravity flow from the blender to a tared product can. The UO_3 product is sampled with a scoop inserted in the falling stream during the transfer. Two samples (one of 30 g and one of 60 g) are taken per product can, one near the beginning and one near the end of the transfer. These samples are put in tared 15-mL glass bottles, sealed in plastic, and transferred to the analytical laboratory. The filled product cans are reweighed to get a net product weight.

In the analytical lab the 60-g sample is reblended, and half reserved as an archive sample. The remaining 30-g is further divided into two 15-g samples, one of which is used for a process control related analysis. The remaining 15-g portion is dissolved in concentrated HNO_3 and diluted to about 500 mL with water. A weight aliquot is taken, spiked with ^{233}U, and mixed. Two 5-drop aliquots of the spiked sample are taken for mass spectrometric analysis.

On some product cans, the 30-g UO_3 sample is also analyzed. This sample is reblended and split. One half becomes an archive sample and the other is treated as just described. If the 30-g sample is not analyzed, the entire 30-g is kept as the archive sample. Generally, both samples are analyzed from each of the first 5 cans of UO_3 product from each denitrator run and from every fifth can thereafter.

Due to the tendancy of UO_3 to absorb water, pains are taken to maintain the integrity of the archive samples. The original, tared sample containers are enclosed in glass vials, sealed with solid polyethylene stoppers, and the stoppers sealed with plastic electrical tape to exclude atmospheric water vapor.

Mass Spectrometry

The solutions obtained by the dissolution of the trioxide in nitric acid are generally clean enough to run in the mass spectrometer without further chemical purification. As before, final product controls are run along with the samples, and a daily instrument calibration is performed. The Quality Control Lab provides a separate set of concentration and ^{235}U limits for use with these samples.

QUALITY CONTROL PROGRAM

The quality control program for the accountability measurements at the ICPP is quite logically centered around defending the IDMS measurement. In order to accomplish this task, the surveillance of every aspect of this measurement is undertaken by the Quality Control Group and much effort is invested in making each step of the measurement traceable to the national measurement system.

The primary indicator of the quality of the uranium measurement is the analysis of uranium standards synthesized from characterized starting materials. The values of the standards (controls) are not known to the laboratories involved in the measurement system. The controls approximate the accountability samples in matrix, uranium concentration, and isotopic composition. The controls are processed at the rate of one per day of plant operation. The controls are rotated by shift, chemical analyst, and mass lab operator. The data generated from the analysis of these controls are monitored with control charts and used to estimate the bias and precision of the various measurements. These precision estimates are used to generate the rejection limits for the comparison between duplicate results mentioned elsewhere in this paper.

The analysis of uranium controls and comparison of duplicate results provide control over the quality of the overall IDMS measurement. We also monitor and control each particular function within this measurement process.

Surveillance of particular functions begins with the preparation and dispensing of the spike aliquots for the IDMS measurement. Spikes are prepared on an individual weight basis from a master solution which is stored in weighed, sealed glass ampoules. The ampoule containing the spike solution is reweighed at the time the ampoule is selected for use. The absence of a significant difference in the two weights eliminates the possibility of concentration of the spike solution caused by a defective seal. The actual process of providing a weight for the individual spikes is performed on an automated weighing system composed of a Mettler HE-20 balance, with associated control units, interfaced to an HP-9810 calculator with printer. This system provides a hardcopy output that displays gross, tare, and net weights and the spike weight in milligrams uranium as calculated from the net spike solution weight and the spike concentration. Advantages of this system include speed, elimination of possible transcription and calculational errors, individualized spike weights, and elimination of suspect spikes. The operation of the balance and associated electronics is checked with the preparation of each ampoule of spike solution by the use of balance weights traceable to the national measurements system.

Also, the various reagents used in the decontamination procedures are extensively monitored for uranium contamination. Large stock solutions of the reagents are prepared, purified as necessary, and analyzed for uranium content. Small lots of these stock solutions are sent to the laboratory and duplicate reagent blanks are run by a low level IDMS procedure. Thereafter, duplicate reagent blanks are processed once per week.

The operation of the remote pipettors and the remote balance is checked with every sample processed by this equipment. This check of the equipment is accomplished by using the known specific gravity (controls) or experimentally determined specific gravity (samples) and the observed sample weight to calculate a volume delivery for the remote pipettors. If the calculated volume delivery differs significantly from the calibrated volume delivery, the sample is flagged and repeated if possible. This check of the remote equipment serves mainly to eliminate gross errors resulting from undetected equipment malfunctions, transposition errors, and other operation errors. In addition, the laboratory personnel responsible for this remote equipment run a daily check on the remote balance and periodically calibrate the remote pipettors.

The operation of the mass spectrometer is checked by the daily analysis of uranium controls, by analyses of duplicate samples, and by the establishment of a daily fractionation correction through the use of primary or secondary standards,

traceable to the national measurements system. The analysis of
standards to determine the fractionation correction is performed
at the beginning of each day shift and, in addition to providing
the fractionation correction, serve as an indicator of any
abrupt change in spectrometer operation. Duplicate samples from
each input measurement batch are processed through the entire
measurement. After the specific gravity homogeneity check, these
samples are assumed to be identical and the difference between
the measured values serves as another indicator of the operation of the entire measurement system. If the difference between the duplicate samples is statistically significant, the
third sample from the accountability tank is measured and a
value is then assigned to the batch.

The operation of the remote densimeters is checked by the
analysis of known bench standards at the rate of one per shift
and bias corrections and precision estimates are generated by
the analysis of unknown controls at the rate of one per month
per analyst using the equipment. The specific gravity values
generated by the remote densimeters are used as the homogeneity
check for the input samples.

EVALUATION

The key features of the accountability measurements system
we have described herein are the capabilities: (a) to make a
large number of measurements, with reasonable accuracy and precision, in the course of a production run: and (b) to produce
simultaneously evidence of the integrity of the data. This is
achieved with a strong reliance on well-established and
generally accepted methodology. It is possible to achieve, with
a much higher cost, somewhat better analysis precision and
accuracy than are generally observed in our system. The extent
to which this would lead to better material balance estimates
can be judged from data presented later in this section. It is
also true that techniques are available, some of which are
described in other papers of this symposium, that can produce
analytical data somewhat faster than those we have chosen to use.

While it is clear that individual components of the measurement system could be improved, we do believe that the system
represents a logical and reasonable approach that has a number
of worthwhile features. In this section we will discuss some of
those features, and then will present data on sources of
uncertainty and their contribution to the uncertainty in the
material balance. In the last section we will discuss projected
refinements of the system.

Sample Throughput

The best illustration of the sample throughput capacity of
the measurement system is derived from the number of mass
spectrometer measurements made during a typical coprocess run.

This represents by far the heaviest load on the system in terms of samples per unit time, as shown in Table II. It is worth mentioning that these figures do not represent maximum throughput of the system as the mass spectrometry laboratory only operates two shifts a day. A 50% increase in sample load could be accommodated by operating a third shift. A second notable point is that the effort devoted to quality control is high. In this case it amounted to almost 23% of the process support workload.

TABLE II. SAMPLE THROUGHPUT

Sample Type		Number
Process Support		1160
Quality Control Program:		
Controls	96	
Reagent Blanks	60	
Mass Spectrometer Calibration	100	
Training	8	
		264
Other Programs and Projects		286
TOTAL		1710

Days of process operation: 104
Samples per 8-hour shift: 7.5

Internal Checks

Several of these have already been mentioned. In general, our philosophy has been that all steps that might easily be challenged are verified as far as possible by some type of internal checking procedure. Examples include confirmation of tank mass measurements by comparing several manometer readings, comparison of input sample specific gravity measurements to assure representative sampling, and comparison of sample weight with volume delivered and specific gravity. In general, such checks have been introduced whenever a number is read by an operator and manually transcribed. The benefits have been greater care in taking such readings, and immediate observation of transcription errors.

Other checks and calibrations serve to ensure against more generalized sources of error. Examples already mentioned include the daily calibration of the mass spectrometer. The long-term variation in the $^{235}U/^{238}U$ ratio, assignable to mass spectrometer operation, has been about 0.13% (one standard deviation). Also included in this category is the comparison of duplicate measurements which is performed on all input and most final product.

Performance
Analytical Measurements

The performance of the analytical methods, as derived from
the analysis of quality control data is shown in Table III. In
general, the analysis precision as derived from quality control

TABLE III

QUALITY CONTROL RESULTS FOR ACCOUNTABILITY CHEMICAL ANALYSES

	95% Confidence Interval, as % of mean	
Sample Type	Concentration	235_U
Aluminum dissolver solution	0.40	0.06
Zirconium dissolver solution*	1.08	0.13
Coprocess solution	0.58	0.08
Electrolytic dissolver solution	0.63	0.12
Final Product	0.57	0.16
Waste streams	40.	----

*Older data may not be representative of current practices.

Data agrees very well with precision estimates made by comparison of duplicate samples. Some of the comparative values are
shown in Table IV.

TABLE IV

COMPARISON OF PRECISION ESTIMATES

	Relative Standard Deviation, %	
Sample Type	From Duplicates	From Quality Control
Aluminum dissolver	0.38	0.40
Zirconium dissolver	0.48	0.52
Coprocess	0.45	0.29
Overall	0.44	0.39

The good agreement of the precision estimates derived by these
two methods lends some confidence in the validity of the quality
control program.

Contribution to Material Balance Calculations

The most important value derived from the accountability measurement system is the "Book-Physical Inventory Difference" (BPID) which is obtained by subtracting the plant holdup and all output terms from the sum of all inputs, thus:

$$BPID = (V_b + S) - (V_e + P + W)$$

where V_b and V_e are the beginning and ending inventories,
S is the input term,
P is the final product output term, and
W is the waste stream output term.
The limit of error (LE) on the BPID (∿ 95% confidence interval) is also important. It indicates whether the BPID differs significantly from zero. The BPID-LE is a combination of all possible sources of measurement uncertainty. The various contributions to a typical BPID-LE are shown as percentages of the total LE in Table V.

TABLE V

BREAKDOWN OF CONTRIBUTIONS TO THE BPID-LE

Source of Uncertainty

Measurement Term	Uranium Analysis	Bulk Weight Measurement	Other	Total
Inventory and Waste	16%	7%	7%	30%
Input	23%	42%	--	64%
Final Product	6%	---	--	6%
Total	44%	49%	7%	100%

Two points are readily apparent: (1), the uncertainty in measurement of the weight of the input solution is the largest single contribution; and (2), the cumulative uncertainties in the various uranium analyses account for nearly half of the LE. The observations suggest the most advantageous directions for future development.

STUDIES IN PROGRESS

It is readily apparent that the greatest potential for improvement of the overall system lies in the area of the solution mass measurement. Unfortunately, this is also the area where it will be most difficult and expensive to make effective improvements since modification of the physical plant itself will be required. Nevertheless, considerable work is under way in this area. Reading the manometers to determine liquid levels in the accountability tanks is currently one of the most easily questioned operations. A direct-reading pressure gauge (Ruska Model DDR-6000) is being installed on one set of accountability tanks for evaluation and will be tested during the next fuel processing campaign. The pressure gauge will be calibrated by comparison to a dead weight gauge (Ruska Model 2465) which is traceable to the national measurements system.

A second area of concern is sampling. Although the aforementioned homogeneity checks and comparisons between duplicate samples provide reasonable assurance of representative sampling, a better understanding of the sampler operation is desireable. We have underway an independent study of mixing in and sampling from accountability tanks, using dye tracers, that is expected to give a better understanding of the sampling/mixing process.

The current sampler design is not one that is easily decontaminated. A new sampler design has been completed, and when installed is expected to result in lower radiation exposures for the operators and to be more conducive to careful sampling.

Even though the uranium measurement is not itself the major source of uncertainty, it is worthwhile to reduce its contribution to the LE as much as economically possible. For example, we have recently considered using a titrimetric method ($\underline{13}$, $\underline{14}$) for measuring total uranium in the UO_3 final product. A several-fold decrease in the uncertainty of that particular uranium concentration would be expected as a result. However, propagation of this source of error through the BPID calculation shows that the maximum improvement in the overall LE would be about 4% relative. In this case, the return is not considered to be cost-efffective in the face of the additional training and quality control effort required. In addition, the cost per analysis would approximately double, since mass spectrometric measurement of ^{235}U would still be required.

A better approach seems to be improvement of the IDMS method performance. To this end, we have been looking at improved separation procedures. We are also extending available information on impurity effects in the IDMS procedure. A new mass spectrometer having greater sensitivity and better reproducibility has been installed and is undergoing performance tests.

Improvement in the hardware used for sample preparation is also in progress. The present remote pipettors use analog circuitry to control volume delivery, resulting in a need for frequent maintenance and calibration. We are designing a microprocessor-controlled pipet that should require less maintenance and be easier to operate. The balances used for remote weighing are converted manual analytical balances which must be operated with remote manipulators. The operator.must read the weight at several feet through a lead glass window. We expect soon to install for evaluation one or more totally remotely controlled balances with direct printout of the sample weight. Not only should this eliminate transcription errors, but it should also reduce mechanical abuse of the balances.

Finally, a new data processing facility is being implemented that will improve the response time of the quality control program. The benefits will be tighter control of performance and quicker recognition of analysis problems.

ACKNOWLEDGEMENT

The successful operation of the accountability measurements system depends on the continual and conscientious support of many people. Some of those most closely involved with the work herein include: G. D. Halverson and A. R. Camarata, Safeguards and Security Office; G. J. Curtis, S. D. Warzel, and N. R. Zack, Quality Control Group; D. R. Trammell, D. N. LeMaire, and the chemists and technicians of the Chemical Analysis Section; and G. W. Webb and the instrument operators of the Heavy Element Mass Spectrometry Group.
This work was performed under U. S. Department of Energy Contract EY-76-C-07-1540.

ABSTRACT

The ICPP has been engaged for 25 years in the recovery of uranium from spent reactor fuels. In concert with the reprocessing activity, an accountability measurements system has been operated throughout the history of the ICPP. The structure and functions of the accountability measurements system are presented. Its performance is evaluated in order to illustrate the relation of analytical methodology to the overall measurements system.

LITERATURE CITED

1. Anon., "INEL", pp. 16-20, Idaho National Engineering Laboratory, U. S. Department of Energy, Idaho Falls, Idaho, 1977.

2. Kratky, O., Leopold, H., and Stabinger, H., Z. Angew. Phys., 27, 273 (1969)

3. Fortsch, E. M., and Wade, M. A., Proc. 22nd Conf. Remote Systems Technology, 1974, 30.

4. Dykes, F. W., Morgan, J. P., and Reider, W. G., "The Remote Analytical Facility Model 'B' Pipetter", IDO-14456, Idaho Operations Office, USAEC, 1958.

5. Maeck, W. J., Booman, G. L., Elliot, M. C., and Rein, J. E., Anal. Chem., 30, 1902 (1958)

6. Shank, R. C., and Crawford, J. M., ed., "Analytical Methods Manual Part I", ICP-1029, Allied Chemical Corp., Idaho Falls, Idaho, 1973.

7.   Echo, M. W., and Morgan, T. D., Anal. Chem., 29, 1953 (1957).

8.   Shank, R. C., ed., "Analytical Methods Manual, Part III", ICP-1031, Allied Chemical Corp., Idaho Falls, Idaho, 1976.

9.   Stevens, C. M., Rev. Sci. Instrum., 24, 148 (1957).

10.   Delmore, J. E., Chem. Instrum., 3, 251 (1972).

11.   Centanni, F. A., Rose, A. M., and DeSesa, M. A., Anal. Chem., 28, 1651 (1956).

12.   Rodden, C. J., ed., "Selected Measurement Methods for Plutonium and Uranium in the Nuclear Fuel Cycle", TID-7029 (2nd ed.), p. 232, USAEC, 1972.

13.   Davies, W., and Gray, W., TRG-Report-716, U. K. Atomic Energy Authority, 1964.

14.   Eberle, A. R., Lerner, M. W., Goldbeck, C. G., and Rodden, C. J., "Titrimetric Determination of Uranium in Product, Fuel, and Scrap Materials after Ferrous Ion Reduction in Phosphoric Acid", NBL-252, New Brunswick Laboratory, New Brunswick, N. J., 1970.

RECEIVED JUNE 30, 1978.

INDEX

INDEX